Springer Series in Optical Sciences

Founding Editor

H. K. V. Lotsch

Volume 247

Springer Series in Optical Sciences is led by Editor-in-Chief William T. Rhodes, Florida Atlantic University, USA, and provides an expanding selection of research monographs in all major areas of optics:

- lasers and quantum optics
- ultrafast phenomena
- optical spectroscopy techniques
- optoelectronics
- information optics
- applied laser technology
- industrial applications and
- synchrotron radiation and X-ray optics
- other topics of contemporary interest.

With this broad coverage of topics the series is useful to research scientists and engineers who need up-to-date reference books.

G. Lozano C. · J. Chacaliaza-Ricaldi ·
J. F. M. dos Santos · E. Marega Jr. · Y. Messaddeq ·
V. A. G. Rivera

A Brief Guide to Optical Spectroscopy of Trivalent Rare Earth Ions in Glasses

Springer

G. Lozano C.
Instituto de Química de São Carlos
Universidade de São Paulo
São Carlos, Brazil

J. F. M. dos Santos
Departamento de Física e Química,
Faculdade de Engenharia
Universidade Estadual Paulista (UNESP)
Ilha Solteira, Brazil

Y. Messaddeq
Centre d'optique, photonique et laser
Université Laval
Quebec, Canada

J. Chacaliaza-Ricaldi
Instituto de Física de São Carlos
Universidade de São Paulo
São Carlos, Brazil

E. Marega Jr.
Instituto de Física de São Carlos
Universidade de São Paulo
São Carlos, Brazil

V. A. G. Rivera
Instituto de Física de São Carlos
Universidade de São Paulo
São Carlos, Brazil

ISSN 0342-4111 ISSN 1556-1534 (electronic)
Springer Series in Optical Sciences
ISBN 978-3-032-18500-6 ISBN 978-3-032-18501-3 (eBook)
https://doi.org/10.1007/978-3-032-18501-3

This Springer imprint is published by the registered company Springer Nature Switzerland AG
The registered company address is: Gewerbestrasse 11, 6330 Cham, Switzerland

To my grandparents, Leo and Clara, my parents, Carmen and Gastón, and my sister Diana. Your influence has shaped my love for books and my passion for learning.

—G. Lozano C.

Modern alchemists do not seek to turn lead into gold, but to turn light into knowledge.

—J. Chacaliaza-Ricaldi

To my parents, Adriana and Casemiro, and my siblings, Rafael and Bruna, for their unwavering support.

—J. F. M. dos Santos

To my parents Antonio and Liliana, thank you for everything.

—V. A. G. Rivera

Preface

Spectroscopic techniques are fundamental to scientific research and technological development, as they allow for the study and application of the interaction between matter and electromagnetic radiation. These methods are true pillars of modern science, and their application in the study of rare-earth ions in glasses—such as those discussed in this book—has impacted both basic research and promising breakthrough technologies.

The field of optical spectroscopy of trivalent rare-earth ions in glasses represents a fascinating intersection between quantum theory and electrodynamics, materials science, photonics, and quantum technologies. In this context, glasses doped with rare-earth ions have become materials for a wide range of applications, including fiber amplifiers, optical amplifiers, sensors, solid-state lasers, optoelectronic device platforms, and quantum memory, just to mention a few examples. Today, rare-earth-doped glasses can be considered a lever for quantum technologies. Their unique optical properties, derived from the shielded 4f electron configurations of rare earths and their interaction with amorphous hosts, make them a valuable subject for both fundamental research and technological innovation.

This book, *A Brief Guide to Optical Spectroscopy of Trivalent Rare Earth Ions in Glasses*, is designed to serve as a comprehensive and accessible resource for students, researchers, and engineers seeking to master the spectroscopic analysis of these materials and apply their knowledge. The book is structured in nine chapters. The text systematically covers key topics such as absorption and emission spectra, decay lifetime dynamics, refractive index, energy transfer processes, the renowned Judd–Ofelt theory, and other complementary quantum-formalism theories. Each chapter covers, in addition to the theoretical foundations, practical methodologies for data interpretation, from experimental results, ensuring that the readers can connect the principles and their interpretations to further their applications.

A distinctive feature of this guide is its emphasis on the interplay between glass composition, local symmetry, and the behavior of rare-earth ions. For example, the analysis of trivalent europium or erbium ions highlights how crystal field distortions enable otherwise forbidden transitions, while the sections on inhomogeneous broadening mechanisms and up-conversion address the challenges and opportunities of quantum technologies. Energy transfer processes between these ions are also discussed, leading to new features such as bandwidth enhancement, white light generation, and optical thermometry. The book also explores cutting-edge advances such as electromagnetically induced transparency in solid-state systems and the role of rare-earth ions in high-power fiber lasers, underscoring their relevance to modern photonics.

The motivation behind this work is twofold: to demystify complex spectroscopic techniques for beginners and to offer a consolidated reference for experienced practitioners. Whether you are investigating optical thermometry, refining glass composition to improve emission efficiency, or exploring rare-earth ions for quantum applications, this book will provide you with the analytical tools and knowledge necessary to advance your research.

We are grateful to the entire scientific community whose pioneering work has shaped this field, and we hope this contribution inspires further exploration and innovation not only in optics and photonics but also in other fields that may complement it.

São Carlos, Brazil V. A. G. Rivera

Acknowledgements

To the São Paulo Research Foundation—FAPESP under the grants (2023/12143-2, 2023/08401-6, 2023/11979-0, 2024/22354-3) and the Conselho Nacional de Desenvolvimento Científico e Tecnológico (CNPq) through the Centro de Pesquisas em Óptica e Fotônica (CePOF)—São Paulo—Brazil (processes 314505/2021–0 and 386305/2024/2), and FAPESP grant 2013/07276-1. This study was financed in part by the Coordenação de Aperfeiçoamento de Pessoal de Nível Superior—Brazil (CAPES)—Finance Code 001.

The authors Aaron, Gaston, Jessica, and Victor are deeply grateful to the São Carlos Institute of Physics for its invaluable role in shaping their academic adventure and for guiding their passion for research.

Contents

Chapter 1
Introduction

Today, we can observe cells at a microscopic scale or admire and study the universe through the lenses of the GRANTECAN Telescope [1]. All of this and much more are possible thanks to researchers' ability to understand the complex properties of glass materials. This fascinating and astonishing development has only been possible due to new advances in the measurement of technical characteristics that allow for a better understanding of the structural, optical, and thermal properties of glasses. Furthermore, over the past 30 years, we have participated in a significant competition to explore new glass compositions that lead to new applications, especially in optics, photonics, telecommunications, and even integrated photonics. One of the keys that enables the use of glass materials in diverse application areas in these fields is their ease of molding into different shapes, from thin, flexible films to microstructure fibers.

In this scenario, trivalent rare-earth-ion (REI)-doped glasses remain the most heavily utilized method in laser glass. The 4f electrons are responsible for the optical emissions and couple weakly to the glass structure. The presence of multiple structural sites in the glass creates a range of dipole environments, which is ideal for engineering broadband REI-doped photonic device materials, which are suitable for laser and amplifier devices. Besides, REI-doped glasses are transparent, which can amplify, with loss reduces, light by the stimulated emission of radiation, i.e., a source of solid-state laser. For a good laser glass performance emission, the identification of the glass composition goes through finding the REI doping concentration ideal and the solubility glass for the REI. Moreover, the performance of the laser glass is also influenced by residual hydroxyl (O–H) groups within the glass structure introduced by conventional melt quenching techniques; hence, it is important to consider purification techniques in the fabrication process of the glass to optimize the laser glass performance [2, 3].

For active devices, high REI doping concentrations are usually required for integrated optical circuits; therefore, glasses with high solubility to REI are required. This is because REI clustering effects become important, leading to low emission

G. Lozano C. et al., *A Brief Guide to Optical Spectroscopy of Trivalent Rare Earth Ions in Glasses*, Springer Series in Optical Sciences 247,
https://doi.org/10.1007/978-3-032-18501-3_1

performance. For example, the low solubility of REI in glasses of silica or chalcogenide due to the mismatch in size, valence, or ion covalency between REI and the constituents of the glass network is an issue to overcome, but other glasses overcome such an issue [4–6].

The development of optical telecommunication is based on the growth of technologies such as fiber fabrication and laser diodes. However, the insatiable need for high-capacity communications results in a permanent reconsideration of the glasses to employ, as well as a new manner of transmission of information. In this context, optical fibers are most often used to transmit information between two points with high velocity of transmission and low loss, and this can also be used as an active medium for amplification purposes and laser development. In these cases, REI-doped fibers are used for their light emission properties (gain and amplification). The dense wavelength division multiplexing (DWDM) technique was developed to meet the demand for multichannel transmission, which was until now supported by Er^{3+}-doped fiber amplifiers (EDFAs). The installation of EDFAs in DWDM networks has led to increases in the speed of data communication, and over the last 25 years, the speed has increased from 100 Mb/s to more than 5 terabits (Tb/s). However, as the data speed approaches several Tb/s, there is a bandwidth limitation in the traditional EDFA [7]. With the development and installation of special optical fibers, the opportunity for extended bandwidth can get over the C-, L-, and U-bands via co-doping of the REIs, which have efficient optical transitions inside and outside these two bands. In other words, the gain per unit length of the fiber depends on the concentrations of dispersed REI in the fiber. Those ions act as the emission centers in the medium for light amplification. Therefore, to design the emission spectra of REI-doped glass and consequently of optical fiber, it is of vital importance to understand the relationship between the local ligand field and the optical properties of specific 4f levels (radiative transition probability, decay probability, and quantum efficiency of amplification transitions).

Plasmonics may join electronics and photonics at the nanoscale by providing an actual solution to the problems of integration and data transfer rate to the nanoscale. Such a new approximation is called Nanophotonics and is dedicated to nanoopto-electronic components, offering extended bandwidths and high processing speeds. However, it remains a challenge to use plasmonic circuits in on-chip configurations due to the required precise control over the plasmonic modes and their properties, particularly those related to their loss [8–10]. In this context, rare-earth-ion (REI)-doped glasses can be employed as substrates. Furthermore, hybrid materials that combine plasmonic nanostructures with a gain medium—such as glass thin films—have demonstrated significant potential for applications in photonics, fuel cells, and other fields, owing to their unique optical properties [10–12].

Finally, the increase in the demand for new temperature sensors gave rise to REI-doped glasses as a new material for optical thermometry due to their temperature dependence on the emission properties of RE-doped materials.

From the above analysis, it is evident that REI-doped glasses play a pivotal role in driving technological advancements and fostering innovative products for global markets. However, before these materials can be integrated into functional devices, a systematic production process must be followed, comprising:

1. Planning and design of the desired device.
2. Fabrication of samples meeting target specifications.
3. Characterization of samples to evaluate key properties.
4. Device simulations incorporating experimental parameters.
5. Prototype manufacturing and validation based on simulation results.

Steps 3 (Characterization) and 5 (Validation) are particularly critical, as they require comprehensive optical and spectroscopic analyses to ensure material performance. In this context, our book serves as an essential resource—not only by detailing spectroscopic techniques but also by providing methodologies for data interpretation and processing to derive meaningful conclusions.

Spectroscopic characterization techniques have historically contributed to groundbreaking scientific achievements, some of which have been recognized with Nobel Prizes. Notable examples include

- William Henry Bragg and William Lawrence Bragg (Nobel Prize in Physics, 1915), who formulated Bragg's law, the foundation of X-ray crystallography—a spectroscopic method to determine material structures.
- Ahmed Zewail (Nobel Prize in Chemistry, 1999), who pioneered femtosecond spectroscopy, enabling real-time observation of molecular dynamics during chemical reactions.
- Eric Betzig, Stefan W. Hell, and William E. Moerner (Nobel Prize in Chemistry, 2014), whose work on super-resolution fluorescence microscopy revolutionized imaging by surpassing the diffraction limits of conventional optical microscopy.

These techniques, among many others, provide fundamental insights into material properties—particularly for REI-doped glasses—enabling researchers to understand their intrinsic behavior, optimize their composition for specific applications, and tailor their performance to meet demands unmet by conventional materials. Thus, spectroscopic characterization holds dual significance: it advances fundamental scientific knowledge and also guiding the practical development of REI-doped glasses for cutting-edge applications.

The almost unlimited versatility of working with glasses that are transparent in various spectral regions or others capable of covering almost the entire electromagnetic spectrum makes them a fascinating material to exploit for diverse applications. Added to this is their ability to accommodate REI, sizes, and charges (limited only by the glass network), which places REI-doped glasses at the center of research for thousands of scientists around the world in the areas of optics, photonics, optical telecommunications, integrated photonics, and quantum communications.

As we shall see in the book, if the desired property is not found for a particular combination of ions, they can be exchanged for others until the desired result (or at least an improved one) is achieved. Sometimes, this can be achieved by substituting co-doped REIs or modifiers for the glass lattice, or even the host material itself, but it is advisable to understand the effects of the glass lattice and its ability to work with REIs in advance to obtain an optimal result.

The world of REI-doped glasses is larger than we imagine, not only because a new glass is synthesized almost daily but also because the resurgence of already known structures accelerates the discovery of new ones and stimulates research in areas with fascinating technological applications and provides business niches. However, this generates a mixture of frustration (too much to read) and enthusiasm (too many interesting things to read and do). What hidden properties persist in the thousands of compounds already manufactured? After decades of research, the field of REI-doped glasses is far from fully developed, and therefore great opportunities for new technological applications remain, an end that is still far from being reached [3, 13, 14].

References

1. S.J. Ikhmayies (ed.), *Advances in Glass Research* (Springer International Publishing, Cham, 2023)
2. D.H. Spadoti, I.C. Pinto, V.A.G. Rivera, Y. Messaddeq, Opt. Express **33**, 10069 (2025)
3. R. El-Mallawany, *Tellurite Glasses Handbook, Physical Properties and Data* (2002)
4. Y.D. Huang, M. Mortier, F. Auzel, Opt. Mater. **15**, 243 (2001)
5. J. Heo, J. Non-Cryst. Solids **326–327**, 410 (2003)
6. J. Heo, W.J. Chung, in *Chalcogenide Glasses*, ed. by J.-L. Adam, X. Zhang (Woodhead Publishing, 2014), pp. 347–380
7. S. Tanabe, C. R. Chim. **5**, 815 (2002)
8. V.A.G. Rivera, O.B. Silva, Y. Ledemi, Y. Messaddeq, E. Marega, *Collective Plasmon-Modes in Gain Media: Quantum Emitters and Plasmonic Nanostructures* (Springer International Publishing, Cham, 2015)
9. T. Zhong, J.M. Kindem, J.G. Bartholomew, J. Rochman, I. Craiciu, E. Miyazono, M. Bettinelli, E. Cavalli, V. Verma, S.W. Nam, F. Marsili, M.D. Shaw, A.D. Beyer, A. Faraon, Science **357**, 1392 (2017)
10. G.C. Lozano, V.A.G. Rivera, S. Celaschi, Y. Messaddeq, E. Marega, Mater. Res. Bull. **180**, 113038 (2024)
11. V.A.G. Rivera, F.A. Ferri, M.H. Gehlen, L.A.O. Nunes, E. Marega, Opt. Eng. **57**, 1 (2018)
12. J.Y.A.C. Ricaldi, V.A.G. Rivera, J.L.C. Huamán, E.M. Junior, in *Glass Science—Innovations and Applications*, ed. by H.C. Vasconcelos (IntechOpen, London, 2025)
13. V.A.G. Rivera, D. Manzani (eds.), *Technological Advances in Tellurite Glasses: Properties, Processing, and Applications* (Springer International Publishing, Cham, 2017)
14. R. El-Mallawany, *The Physics of Advanced Optical Materials: Tellurite Glasses* (Springer Nature Switzerland, Cham, 2023)

Chapter 2
Absorption Spectrum

An absorption spectrum is a graphical representation of the amount of electromagnetic radiation absorbed by a material as a function of the wavelength of the incident light. It is a fundamental analytical tool in chemistry, physics, astronomy, and materials science, providing insights into the composition, structure, and energy levels of matter [1]. The absorption spectrum is one of the most fundamental and widely used tools for studying the optical properties of glasses doped with rare-earth ions (REIs) [2]. It describes how a material absorbs electromagnetic radiation as a function of wavelength or photon energy, providing information on its electronic structure, optical transitions, and degree of disorder [3, 4]. In REI-doped glasses, absorption arises predominantly from 4f–4f transitions within trivalent lanthanide ions, which are partially shielded by outer electron shells and therefore retain sharp spectral features even in amorphous hosts [5].

The history of absorption of spectroscopy dates back to the seventeenth century. In 1666, Sir Isaac Newton demonstrated the decomposition of white light into its spectral components using a prism, laying the foundation for the study of light–matter interaction. In 1802, William Wollaston observed dark lines in the solar spectrum, later cataloged by Joseph von Fraunhofer. These Fraunhofer lines played a crucial role in the identification of chemical elements. In 1852, August Beer formulated a quantitative relationship describing absorbance in terms of concentration and optical path length, which, together with earlier contributions by Pierre Bouguer and Johann Heinrich Lambert, formed the basis of the Beer–Lambert law. During the twentieth century, the technique advanced significantly with the development of atomic and molecular absorption spectroscopy, becoming one of the most precise methods for material characterization [6, 7]. In the context of modern photonic materials, especially glasses doped with RE ions, such as Er^{3+}, Nd^{3+}, or Eu^{3+}, absorption spectra serve as a basis for analyzing energy levels; calculating Judd–Ofelt parameters; and designing devices for lasers, amplifiers, and luminescent probes [8].

G. Lozano C. et al., *A Brief Guide to Optical Spectroscopy of Trivalent Rare Earth Ions in Glasses*, Springer Series in Optical Sciences 247,
https://doi.org/10.1007/978-3-032-18501-3_2

2.1 The Beer–Lambert Law

The Beer–Lambert law describes the absorption of light in materials, which is an empirical relationship that relates the attenuation of light to the properties of the material through which the light travels. The origins of the law can be traced back to Pierre Bouguer, who coined the idea that would later be known as the Beer–Lambert law in the summer of 1729 in his *Essay on the gradation of light*, where he defined the amount of light lost when it passes through a certain extent of the atmosphere. This work later served Johann Heinrich Lambert in 1760 in his work *Photometry*, where he noted that the loss of light intensity as it propagated in a medium is directly proportional to the intensity and length of the optical path. Almost a century later, in 1852, August Beer proposed another relationship for light attenuation, where he established that the transmittance of a solution remains constant if the product of concentration and the optical path length remains constant [9, 10].

In this context, the modern derivation of the Beer–Lambert law combines the two propositions and correlates the optical attenuation of a material containing a single attenuating species of uniform concentration with the length of the optical path as follows:

$$A = \alpha l c \tag{2.1}$$

where A is absorbance, l is the length of the optical path expressed in cm, c is the concentration, and α is the absorption coefficient or absorptivity of the attenuating materials, defined as follows:

$$\alpha = \frac{4\pi k}{\lambda} \tag{2.2}$$

where k is the attenuation coefficient and λ is the wavelength. The absorption of light by a material is quantified by its absorption coefficient α, which defines the fraction of power absorbed per unit length of the material. If the beam propagates in the z direction, and the intensity (optical power per unit area) at position z is $I(z)$, then the decrease in intensity over a thickness of dz is given by

$$dI = -\alpha I(z) dz \tag{2.3}$$

This equation can be integrated to obtain the following expression:

$$I(z) = I(0)e^{-\alpha z} \tag{2.4}$$

where I_0 is the optical intensity at $z = 0$. The absorption coefficient is a function of wavelength, so optical materials may or may not absorb certain wavelengths. The absorption of electromagnetic radiation by materials from the UV–visible region

(300–400 nm and 400–800 nm) to the NIR region (800–2000 nm) provides information about the material's structure and electronic level transitions. With this information, the composition of the material one can be identified from the absorption spectrum. These transitions occur when atomic systems in the ground state absorb radiation, causing them to transition to a higher energy excited state. Radiation absorption in the UV and visible range causes electronic and vibrational excitations, while absorption in the NIR range causes rotational and vibrational transitions [11, 12].

UV–Vis absorption measurements are typically performed using a scanning spectrophotometer, which enables the acquisition of high-resolution absorbance spectra in ultraviolet, visible, and near-infrared regions. The experimental arrangement, schematized in Fig. 2.1, consists of the following components: A broadband light source (typically a deuterium lamp for UV and a tungsten–halogen lamp for Vis–NIR) emits radiation that is collimated and dispersed by a diffraction grating. A monochromatic beam is selected by a rotating grating and directed through a slit, defining its spectral resolution and spatial coherence of the selected wavelength. The incident intensity I_0 passes through the sample, and the transmitted intensity I as a function of wavelength is measured by a detector. The detector signal is used to compute the absorbance according to Beer–Lambert's law:

$$\text{Absorbance} = \log_{10}\left(\frac{I_0}{I}\right) \tag{2.5}$$

This absorbance provides a fingerprint of electronic transitions present in the sample, such as the 4f–4f transitions in RE-doped glasses.

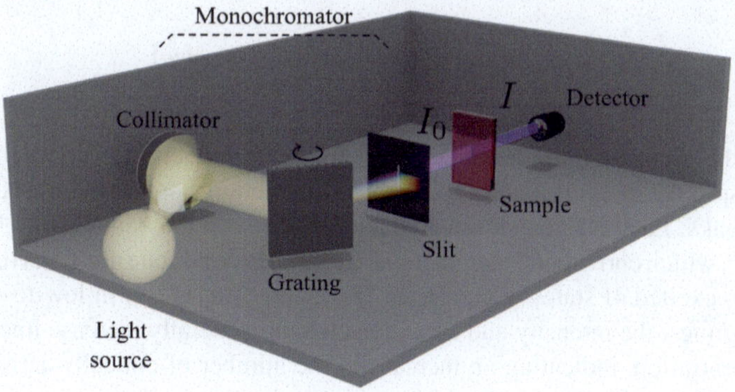

Fig. 2.1 Schematic diagram of a UV–Vis absorption spectrophotometer setup

2.2 Rare-Earth-Doped Glasses Absorption Spectra

The study of absorption spectra in RE-doped glasses provides crucial information about the optical properties and electronic structure of these materials [13–15]. In such glasses, REIs introduce discrete, sharp absorption bands corresponding to intra-4f electronic transitions that are only weakly influenced by the surrounding disordered matrix due to the shielding of the 4f orbitals by outer 5s and 5p shells [16, 17].

The absorption of light in REI-doped glasses arises from electronic transitions between discrete 4f states of the REI^{3+}; the time-dependent perturbation theory in quantum mechanics fundamentally describes this phenomenon. According to Fermi's golden rule, the transition rate between an initial state $|i\rangle$ and a final state $|f\rangle$ under a perturbation H' (due to interaction with the electromagnetic field) is

$$W_{i \to f} = \frac{2\pi}{\hbar} \left| \langle f | H' | i \rangle \right|^2 \rho(E_f) \tag{2.6}$$

where $\rho(E_f)$ is the density of final states and $H' = -\vec{\mu} \cdot \vec{E}$, where $\vec{\mu}$ is the electric dipole operator and $\vec{E}$ is the electric field vector [18]. The matrix element $|\langle f | \vec{\mu} | i \rangle|^2$ encapsulated the transition dipole moment, dictating selection rules and transition probabilities. In REI, 4f orbitals are shielded by outer 5s and 5p shells, rendering 4f–4f transitions nominally dipole-forbidden. However, in a host matrix, the crystal field perturbs the wavefunctions, mixing states with opposite parity, thus enabling weak but finite dipole transitions [19]. The absorption coefficient $\alpha(\omega)$ becomes proportional to the sum over these transitions weighted by their respective probabilities:

$$\alpha(\omega) \propto \sum_{i,f} |\langle f | \vec{\mu} | i \rangle|^2 \delta(E_f - E_i - \hbar\omega) \tag{2.7}$$

Thus, the fine structure seen in absorption spectra reflects the variety of discrete 4f levels split by spin–orbit and crystal–field interactions, appearing as sharp, well-defined peaks. Each RE ion shows characteristic absorption bands in the UV–Vis–NIR range, which correspond to transitions from the ground state multiplet (e.g., $^4I_{15/2}$ in Er^{3+}) to excited 4f states (e.g., $^2H_{11/2}$, $^4S_{3/2}$, $^4F_{9/2}$, etc.) [20]. In low-to-moderate dopant regimes, the intensity and area of each band generally increase linearly with REI concentration, indicating an increase in the number of optically active centers [21].

Measurements of the absorption spectrum allow the identification of the energy levels and specific transitions of each REI. They also allow for the quantification of the effects of dopant concentration: studying how the area or intensity of absorption peaks scales with the RE content helps reveal clustering, extinction, or uniform scattering of dopant ions. It is worth mentioning that at higher RE contents, deviations from the linear absorption scale may indicate energy transfer or clustering processes [22].

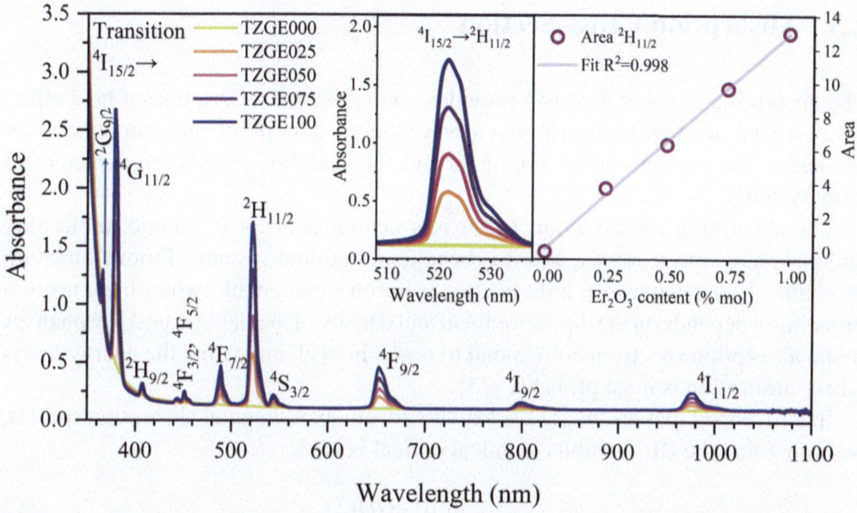

Fig. 2.2 UV–Vis absorption spectra of Er^{3+}-doped tellurite–zinc–sodium glasses. The insets present the $^4I_{15/2} \rightarrow {}^2H_{11/2}$ absorption band and the linear correlation between Er_2O_3 concentration and the integrated area under the $^2H_{11/2}$ band

Another difficulty arises when matrix absorption tails can overlap RE transitions, especially near the UV edges. On the other hand, in multi-doped systems (e.g., Er^{3+}/Tm^{3+} co-doping), absorption bands may overlap, complicating the analysis [23].

Figure 2.2 shows the UV–Vis absorption spectra of glasses with the base composition (70–x)% TeO_2–20% ZnO–10% Na_2O doped with different concentrations of Er_2O_3 (x = 0, 0.25, 0.50, 0.75, and 1.00 mol%). Multiple characteristic bands are observed corresponding to internal f–f transitions of the Er^{3+} ion, labeled in the spectrum and attributed to well-known transitions from the ground level 4I15/2 to excited levels such as 4G9/2, 2H9/2, 4F3/2, 4F5/2, 4F7/2, 2H11/2, 4S3/2, 4F9/2, 4I9/2, and 4I11/2.

The inset highlights the band around 520–530 nm corresponding to the $^4I_{15/2} \rightarrow {}^2H_{11/2}$ transition, showing a systematic increase in both the maximum absorbance and the integrated band area as the Er_2O_3 concentration increases. The plot on the right confirms the linear relationship between the Er_2O_3 content and the $^2H_{11/2}$ band area, with an excellent correlation (R^2 = 0.998). This behavior indicates that the progressive increase in Er^{3+} in the glass matrix leads to a higher density of optically active centers responsible for absorption in this band, without evidence of saturation or significant clustering effects in the studied range. These results suggest that progressive Er^{3+} doping effectively increases the density of absorbing states and the excitation efficiency of intra-band transitions, which is crucial for applications in active optical devices such as amplifiers and Er^{3+}-doped fiber lasers [24].

2.3 Absorption Cross-Section

The absorption cross-section (σ) provides a microscopic description of how effectively a REI absorbs light at a given wavelength. It is particularly important for evaluating the gain in optical amplifiers and for modeling population dynamics in laser systems.

The absorption cross-section (σ_{abs}) is a central concept in atomic, molecular, and gas-phase spectroscopy, and it is beneficial in dilute systems. From a quantum mechanical perspective, σ_{abs} is derived using Fermi's golden rule, where the transition probability depends on the dipole moment and density of available states. Resonances in the absorption spectrum correspond to peaks in $\sigma(v)$, indicating the energy levels where absorption is most probable [25].

In REI-doped glasses, $\sigma_{abs}(\lambda)$ is extracted from experimental absorption spectra, assuming that the REI exhibits identical optical behavior:

$$\sigma_{abs} = \frac{\ln 10 \cdot OD(\lambda)}{N \cdot d} \tag{2.8}$$

where N (cm^{-3}) represents the homogeneous RE ion density, d the sample thickness, and $OD(\lambda) = -\ln(T_{out}(\lambda)/T_{in}(\lambda))$ [26]. The cross-section carries key information about the probability of electronic transitions and is influenced by local symmetry, phonon coupling, and inhomogeneous broadening due to the amorphous nature of the glass.

In many practical systems, only the absorption spectrum is experimentally accessible, while the emission cross-section $\sigma_{em}(\lambda)$ is required to model the gain behavior [27]. To overcome this, McCumber's theory provides a thermodynamically consistent relation between the absorption and emission cross-sections. The theory assumes thermal equilibrium between the sublevels of the ground and excited manifolds and relates the emission and absorption cross-sections as follows:

$$\sigma_{em}(\lambda) = \sigma_{abs}(\lambda) \cdot \exp\left(\frac{\varepsilon - hv}{k_B T}\right) \tag{2.9}$$

where $\sigma_{em}(\lambda)$ and $\sigma_{abs}(\lambda)$ are the emission and absorption cross-sections, ε is the so-called zero-line energy, corresponding to the energy difference between the lowest Stark level of the upper manifold and the lowest Stark level of the lower manifold, hv is the photon energy, k_B is the Boltzmann constant, and T is the absolute temperature [28, 29]. In glasses doped with Er^{3+} (e.g., for 1550 nm telecommunications amplifiers), Yb^{3+} (980–1030 nm lasers), or Nd^{3+} (1064 nm emission), McCumber theory is widely used to derive emission cross-sections from absorption measurements, model the gain spectrum of optical amplifiers, and predict the saturation intensity and population inversion threshold [30–32]. A more in-depth McCumbers's theory will be presented in Chap. 3.

2.4 Tauc Model

In non-crystalline materials such as glasses, the optical bandgap E_g is a key parameter that describes the energy range where no allowed electronic states exist between the valence and conduction bands [33, 34]. The term bandgap refers to the energy range in which no electronic states exist, preventing electrons from transitioning freely between the valence and conduction bands [35]. This gap inhibits free electronic transitions under thermal excitation and determines the threshold energy required for optical absorption. In contrast to crystalline materials, where electronic band structures are well defined due to long-range periodicity, amorphous systems exhibit broadened and localized energy states near the band edges, making the precise definition of the bandgap more complex.

The optical bandgap of glasses is typically determined using UV–Vis–NIR absorption spectroscopy, following the model proposed by Tauc and Menth (1972) [36]. According to this model, the absorption coefficient α is related to the photon energy $h\nu$ as follows:

$$(\alpha h\nu)^{1/r} = A\left(h\nu - E_g\right) \tag{2.10}$$

where A is a proportionality constant related to the probability of electronic transitions, E_g represents the optical absorption edge, and r is a real number that depends on the type of band-to-band transition [37]. The values of r are presented in Table 2.1.

Electronic transitions in materials can be classified as either direct or indirect, depending on the relative alignment of the conduction and valence band extrema in the k-space momentum:

- In direct transitions, the maximum of the valence band and the minimum of the conduction band occur at the same k-value in the reciprocal space. An electron can absorb a photon and be excited directly without additional momentum transfer (Fig. 2.3a).
- In indirect transitions, the band extrema are misaligned in k-space. As a result, an additional interaction, typically with phonons (quantized lattice vibrations), is required to conserve momentum during the transition (Fig. 2.3b).

To determine the optical bandgap energy E_g of RE-doped glasses using the Tauc method, the absorption edge in UV–Vis spectra. The method involves plotting the absorption function $(\alpha h\nu)^{1/r}$ versus $h\nu$. When a linear region is identified in the plot, a linear fit is made on that region. The Tauc relation can be expressed as a linear

Table 2.1 Type of bandgap transition

Type of transition	Allowed transition (r)	Forbidden transition (r)
Direct	1/2	3/2
Indirect	2	3

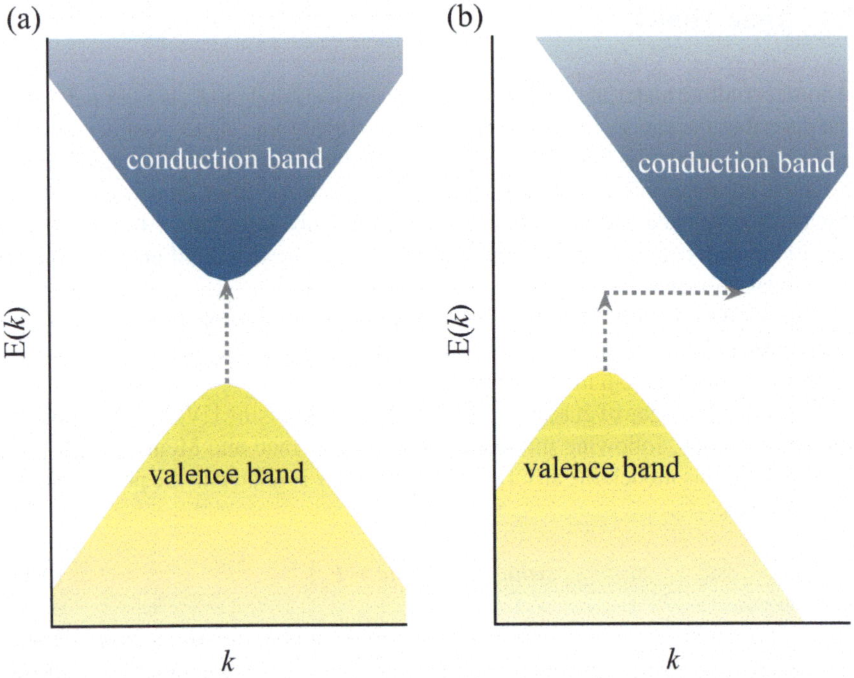

Fig. 2.3 Diagram of band-to-band absorption transitions. **a** Direct and **b** indirect bandgap

equation of the form $y = a + bx$. By extrapolating this line to the energy axis $y = 0$, the intercept gives the value of the optical bandgap $0 = a + bx_0 \rightarrow x_0 = -a/b$.

This value x_0 corresponds to E_g, the photon energy at which the material begins to absorb significantly due to inter-band electronic transitions. Thus, the optical bandgap is graphically extracted as the intersection of the linear fits with the photon energy axis, giving an estimate of E_g in glasses. In Fig. 2.4, a Tauc plot is observed corresponding to the start of the optical absorption edge. A linear fit was applied to this region, and the optical bandgap E_g is obtained by extrapolating the fit to the energy axis (i.e., where $(\alpha h\nu)^{1/r} = 0$). In the example shown, mathematically, the x_0 intercept (or E_g) is obtained from the linear fit with parameters $m = 1.636$ and $b = -5.271$, resulting in $x_0 = -\left(\frac{-5.271}{1.636}\right) = 3.221$ eV. This value represents the optical bandgap of the glass, which reflects the minimum energy required to promote an electron from the valence band to the conduction band, considering both structural disorder and transition type.

In RE-doped glasses, the optical bandgap is sensitive to both the host matrix composition and the dopant concentration, and it can affect non-radiative relaxation processes and multiphonon decay rates [38–40]. Incorporation of REI can influence the bandgap through several mechanisms such as structural modification of the glass network [33, 41, 42]. In contrast, controlling the bandgap by optimizing glass composition and dopant levels allows better tuning of the optical response of the material,

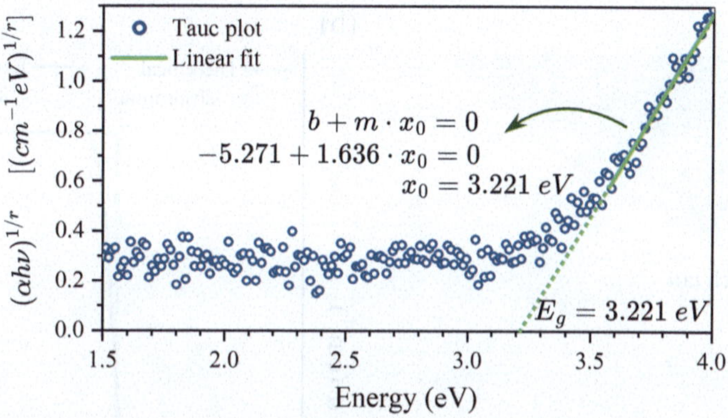

Fig. 2.4 Tauc plot for determining the optical bandgap E_g in glass. The linear fit to the absorption edge produces an extrapolated intercept at x_0, corresponding to the optical bandgap energy

particularly in applications such as optical amplifiers, up-conversion phosphors, and photonic devices [43].

2.5 Urbach Energy

In non-crystalline materials, the absorption coefficient α exhibits an exponential dependence on photon energy near the optical bandgap [44]. This absorption region, known as the Urbach tail, occurs below the Tauc absorption edge, where photon energies are insufficient to induce direct or indirect inter-band transitions. Figure 2.5a represents the typical profile of the electronic density of states (DOS) in a non-crystalline material (such as REIs-doped glasses). It identifies the valence band (in yellow) and the conduction band (in blue), separated by the bandgap. However, unlike ideal crystalline materials, this representation incorporates Urbach tails, which introduce additional electronic structure within the bandgap because of structural disorder. Urbach tails are exponential extensions of the density of states of the bandgap, caused by structural disorder of the amorphous structure, electron–phonon interactions, and the presence of local defects or electrostatic potential fluctuations [45].

Figure 2.5b compares theoretical (green curve) and experimental (purple curve) optical absorption as a function of photon energy. In an ideal crystalline material, absorption exhibits an abrupt edge: it remains nearly zero below the bandgap and rises sharply once the threshold energy is reached. This idealized behavior is represented by the step-like theoretical curve, marking the onset of inter-band electronic transitions [46]. In contrast, the experimental curve displays a gradual slope at the absorption edge, extending into sub-bandgap energies. This behavior results from localized states within the bandgap, induced by structural disorder, atomic defects, or electron–phonon interactions [47]. The deviation from the ideal response forms

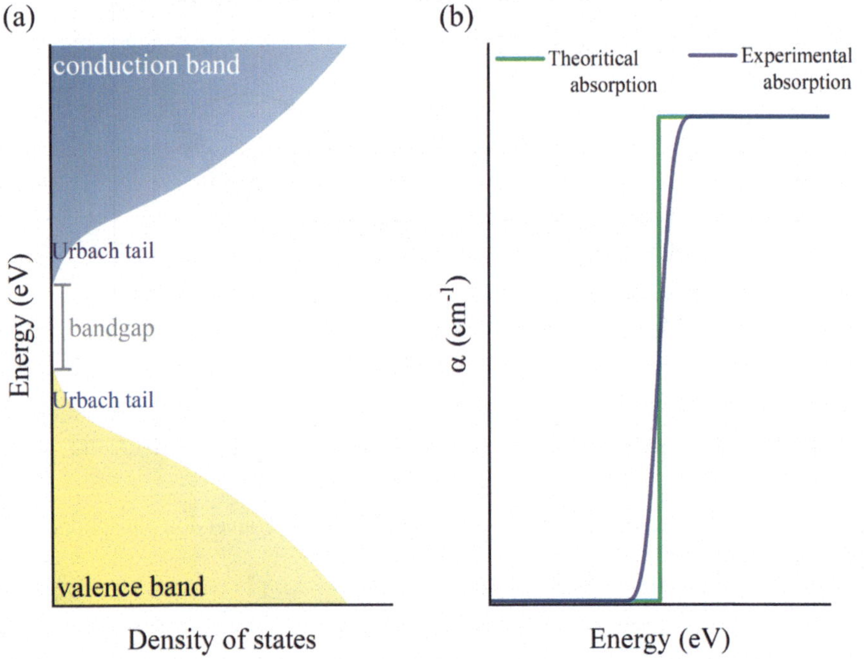

Fig. 2.5 Comparative graph of **a** Urbach tails and **b** theoretical and experimental absorption

the Urbach tail, characterized by an exponential dependence of the absorption coefficient on the photon energy in this region. This discrepancy is not a measurement artifact, but a fundamental feature of disordered materials such as RE-doped glasses. Analyzing this tail yields the Urbach energy as a key parameter for quantifying structural disorder.

This behavior is reflected optically in the region near the absorption edge, where the absorption coefficient follows an exponential dependence known as Urbach's law:

$$\alpha = \alpha_0 \exp\left(\frac{h\nu}{E_U}\right) \tag{2.11}$$

where $\alpha(h\nu)$ is the optical absorption coefficient, E_0 is the optical absorption edge energy (or Tauc energy), and E_U is the Urbach energy, which quantifies the degree of structural disorder of the system [48]. By applying the natural logarithm to both sides of the equation, it can be linear as follows:

$$\ln \alpha = \ln \alpha_0 + \frac{1}{E_U} h\nu \tag{2.12}$$

This expression is mathematically equivalent to the linear equation $y = mx + b$, where the slope $m = 1/E_U$ and the intercept $b = \ln \ln \alpha_0$. Therefore, the Urbach energy can be directly obtained from the inverse of the slope of the linear region in a plot of $\ln \ln \alpha$ versus $h\nu$. This linear region corresponds to the exponential tail of the absorption spectrum and reflects the degree of structural disorder in the glass matrix. A larger E_U value indicates a higher degree of disorder, making this parameter essential for quantifying the optical quality and electronic disorder in amorphous and doped glass materials.

Figure 2.6 shows the graphical representation used to determine the Urbach energy from the optical absorption spectrum of a doped glass. In this analysis, the natural logarithm of the absorption coefficient $\ln \ln \alpha$ is plotted as a function of the photon energy $h\nu$, highlighting the sub-bandgap region where the behavior is exponential. This linear region is associated with transitions assisted by structural disorder and thermal vibrations. The slope of this line is inversely related to the Urbach energy, such that $E_U = 1/m$, where m is the slope of the linear fit. In the example shown, the obtained value is $E_U = 0.358$ eV, which indicates a moderate degree of disorder in the glass lattice of the analyzed material. This parameter is crucial to evaluate the optical quality of the glass and its applicability in photonic devices.

Care must be taken to select the appropriate energy range for fitting—sufficiently below the band edge to avoid overlap with inter-band transitions, but above the noise level. The obtained value of Urbach energy serves as a sensitive probe for the degree of structural and compositional disorder in the material. Higher Urbach energies indicate greater disorder, often associated with the presence of non-bridged oxygens (NBOs), defects, or dopant-induced distortions in the glass matrix [49, 50]. This method is widely applied in the characterization of amorphous semiconductors, glasses doped with RE ions, and other optical materials where band-tail states play a significant role in the absorption mechanism.

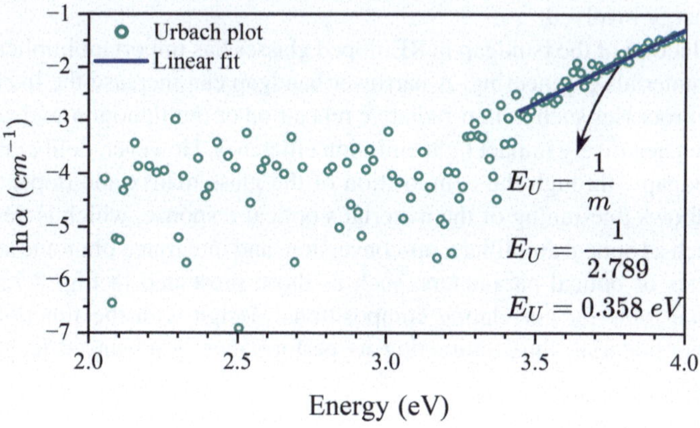

Fig. 2.6 The plot of $\ln \alpha$ versus photon energy $h\nu$ reveals a linear region in the sub-bandgap regime, from which the Urbach energy E_U is extracted as the inverse of the slope

2.6 Influence of Rare-Earth Ions on the Optical Bandgap Energy

Glasses doped with REI are versatile in photonic applications due to their characteristic optical transitions and chemical stability. However, beyond their function as emitting centers, REI significantly influences the structure and optical properties of the host glass. The optical bandgap and Urbach energy, in particular, is sensitive to both matrix composition and dopant concentration, and its modification has direct implications for non-radiative relaxation processes and multiphonon decay rates.

2.6.1 Effects on Direct and Indirect Bandgaps

Figure 2.7 presents the optical bandgap and Urbach energy analysis for three tellurite glass compositions: the base matrix of (60-x–y)% TeO_2–20% ZnO–10% GeO_2–10% Na_2O (TZGN), the Er^3-doped (x%) sample (TZGN_Er), and the Er^{3+}/Yb^{3+}-doped ($x + y$%) sample (TZGN_Er_Yb). Figure 2.7a shows the Tauc plot for direct transitions, where $(\alpha h\nu)^2$ is plotted as a function of photon energy. Linear extrapolation of the absorption edge allows the direct bandgap values to be determined. These results indicate a slight decrease in the bandgap with progressive doping. The introduction of Er^{3+} and Yb^{3+} induces structural modifications in the glass lattice, possibly through the creation of oxygen vacancies or the weakening of bonds in the matrix, generating localized states that reduce the energy required for electronic excitation. Figure 2.7b presents $(\alpha h\nu)^{1/2}$ that corresponds to the Tauc model for indirect transitions. Although the values are lower than those for the direct transition, a similar trend is observed. This confirms that the presence of RE facilitates indirect optical absorption mechanisms, which can favor non-radiative processes, depending on the phonon energy involved.

The reduction of the bandgap in RE-doped glasses has important implications for photonic materials engineering. A narrower bandgap can increase the likelihood of undesired processes such as non-radiative relaxation or multiphonon-assisted transitions, which negatively impact light emission efficiency. However, deliberate control of the bandgap—through the composition of the glass matrix and dopant concentration—allows fine-tuning of the material's optical response, which is desirable in devices such as optical amplifiers, up-conversion, and integrated photonic structures. The analysis of optical parameters, such as those presented in Fig. 2.7, provides fundamental tools for correlating compositional design with the functional properties of the material, thus optimizing its performance in advanced technological applications.

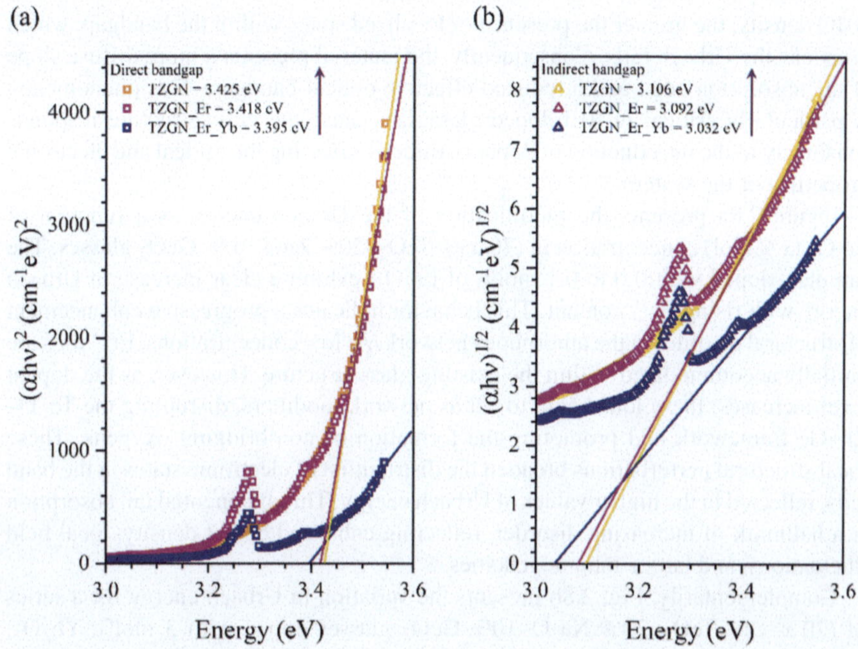

Fig. 2.7 Tauc plots for **a** direct and **b** indirect optical transitions in TZGN glasses doped with Er^{3+} and Er^{3+}/Yb^{3+}

2.6.2 *Effects on Urbach Energy*

As discussed previously, the introduction of RE ions into glasses significantly modifies their structural and optical properties, especially due to the localized 4f electronic states and their weak screening from the surrounding matrix. An interesting example to address is tellurite oxide-based glasses, which are already known for their low phonon energies and high refractive indices. Doping with trivalent RE ions such as Er^{3+}, Yb^{3+}, and Tm^{3+} opens up a wide range of functionalities for photonic applications, including near-infrared amplification, up-conversion luminescence, white light emission, and thermometric sensing. However, in addition to their optical activity, REIs also influence the structural disorder of the host glass, which can be probed through Urbach energy measurements.

In addition to introducing characteristic intra-band electronic levels, doping with REIs also causes significant structural modifications in the glass lattice. One of the most relevant consequences of this process is the generation of NBOs which result from the breaking of covalent bonds between network formers (such as Te–O–Te bonds). As the dopant concentration increases, these modifying ions occupy interstitial positions or replace structural units, breaking bonds and favoring the formation of NBOs. This structural alteration increases the topological disorder of the matrix, which is optically reflected in an increase in the Urbach energy. The higher the

NBO density, the greater the presence of localized states within the bandgap, which broadens the Urbach tails. Consequently, the material presents a more diffuse slope at the absorption edge and a reduced effective optical bandgap. This phenomenon is particularly critical in REI-doped glasses, because the glass structure responds sensitively to the introduction of dopants, directly affecting the optical and electronic properties of the system.

Figure 2.8a presents the modification of the Urbach energy as a function of Er_2O_3 (x %mol) concentration in (70-x)% TeO_2–20% ZnO–10% GeO_2 glasses. The samples, doped with 0.0 to 0.1 mol% of Er_2O_3, exhibit a clear increase in Urbach energy with rising Er^{3+} content. This behavior indicates a progressive enhancement of structural disorder in the amorphous network. At low concentrations, Er^{3+} ions are partially accommodated within the existing glass structure. However, as the dopant level increases, these ions begin to act as network modifiers, disrupting the Te–O–Zn–Ge framework and promoting the formation of non-bridging oxygens. These local structural perturbations broaden the distribution of electronic states in the band tails, reflected in the higher values of Urbach energy. This exponential tail absorption is a hallmark of increasing disorder, reflecting enhanced defect density, local field fluctuations, and lattice inhomogeneities.

Complementarily, Fig. 2.8b presents the variation in Urbach energy for a series of (70-x–y)% TeO_2–20% Na_2O–10% GeO_2 glasses doped with 3 mol% Yb_2O_3,

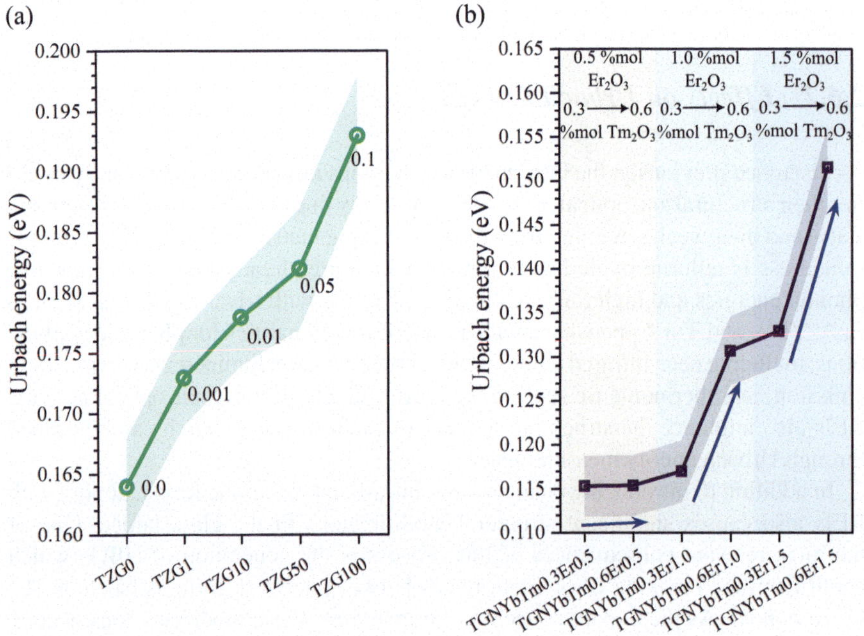

Fig. 2.8 Variation of the Urbach energy as a % molar concentration function of **a** Er^{3+} in TeO_2–ZnO–GeO_2 and **b** $Tm^{3+}/Er^{3+}/Yb^{3+}$ in TeO_2–GeO_2–Na_2O

where the concentrations of (*x* %mol) Er_2O_3 and (*y* %mol) Tm_2O_3 were systematically altered. The graph is organized into three distinct regions highlighted with background shading: yellow for 0.5 mol% Er_2O_3, green for 1.0 mol%, and blue for 1.5 mol%. Within each of these Er^{3+} concentration regimes, the Tm_2O_3 content was varied from 0.3 to 0.6 mol%. In all three cases, an increase in Tm^{3+} concentration leads to a consistent rise in the Urbach energy, indicating enhanced structural disorder. This trend is also observed across the Er_2O_3 series: as the Er^{3+} content increases from 0.5 to 1.5 mol%, the overall Urbach energy also rises, regardless of the fixed Tm^{3+} level.

This behavior reflects the cumulative effect of REIs in the glass lattice. Both Er^{3+} and Tm^{3+} are large, heavy ions with high field strengths, which disrupt the intermediate order of the TeO_2 and GeO_2 units that form the lattice. Their incorporation can lead to enhanced NBO formation, distortions in the local coordination environment, and the introduction of defect states within the optical bandgap. These localized states facilitate tail absorption, broadening the Urbach region. Furthermore, the co-doping strategy can have a synergistic effect on the glass matrix. Although each REI can cause moderate distortion on its own, their combined presence indicates lattice disruption, as evidenced by the larger increase in Urbach energy at high concentrations of Er^{3+} and Tm^{3+}. Furthermore, energy transfer processes between Yb^{3+}, Er^{3+}, and Tm^{3+} can indirectly contribute to shifts in the absorption edge, possibly through phonon-assisted processes or local field modifications.

Therefore, UV–Vis–NIR absorption spectroscopy has proven to be a powerful and versatile tool for probing the electronic structure, optical transitions, and degree of structural disorder in RE-doped glasses. Through the analysis of the absorption edges and the characteristic 4f–4f transitions, key optical parameters such as direct and indirect bandgap energies and Urbach energy can be extracted. These parameters offer valuable insight into the local environment of dopant ions and the integrity of the host matrix.

The results analyzed in this chapter demonstrate that even small additions of RE ions can significantly alter the structural and electronic behavior of glasses. Progressive doping results in a reduction of the optical bandgap and an increase in the Urbach energy, indicating higher localized densities of states and greater structural disorder. In co-doped systems, these effects are amplified due to the synergistic distortion introduced by multiple RE species, often resulting in broader absorption tails and enhanced non-radiative relaxation pathways. These modifications have profound implications for photonic applications. Although increased disorder can enhance processes such as broadband emission or up-conversion, it can also reduce emission efficiency and optical transparency. Therefore, it is essential to balance the beneficial and detrimental effects of RE-induced disorder when designing glass compositions for specific functionalities such as optical amplifiers, lasers, sensors, and nonlinear optical devices.

In the future, one of the main challenges lies in the precise control of REI distribution within the glass matrix, particularly at higher doping levels where clustering or phase separation may occur. Advanced synthesis techniques, combined with in situ spectroscopic diagnostics, will be essential to overcome these limitations. Moreover, there is a growing need to correlate structural disorder with dynamic optical behavior, such as lifetime shortening, energy transfer rates, or thermal quenching. Finally, the development of multifunctional and tunable glass systems, capable of supporting multiple rare-earth ions without compromising optical quality, remains a key goal. Such systems will be pivotal in next-generation technologies for quantum photonics, integrated optics, and energy-efficient light sources.

References

1. J.M. Hollas, Modern spectroscopy (2013), https://www.wiley.com/en-us/Modern+Spectroscopy%2C+4th+Edition-p-9780470844168. Accessed 29 Jul 2025
2. B. Karmakar, Fundamentals of glass and glass nanocomposites, in *Glass Nanocomposites: Synthesis, Properties and Applications* (2016), pp. 3–53. https://doi.org/10.1016/B978-0-323-39309-6.00001-8
3. S. Jackson et al., Spectroscopy of the rare-earth-ion transitions in fluoride glasses, in *Mid-Infrared Fiber Photonics: Glass Materials, Fiber Fabrication and Processing, Laser and Nonlinear Sources* (2022), pp. 333–399. https://doi.org/10.1016/B978-0-12-818017-4.00011-2
4. B. Karmakar, Functionalization of glasses by incorporation of semiconductor, in *Functional Glasses and Glass-Ceramics* (2017), pp. 235–252. https://doi.org/10.1016/B978-0-12-805056-9.00006-4
5. R. Shi, A.V. Mudring, Phonon-mediated nonradiative relaxation in Ln^{3+}-doped luminescent nanocrystals. ACS Mater. Lett. **4**(10), 1882–1903 (2022). https://doi.org/10.1021/ACSMATERIALSLETT.2C00595/ASSET/IMAGES/LARGE/TZ2C00595_0002.JPEG
6. P. Atkins, J. de Paula, J. Keeler, *Atkins' Physical Chemistry* (2022). https://doi.org/10.1093/HESC/9780198847816.001.0001
7. I.N. Levine, *Physical Chemistry*, vol. 6 (2009)
8. J.G. Solé, L.E. Bausá, D. Jaque, *An Introduction to the Optical Spectroscopy of Inorganic Solids* (2005), pp. 1–283. https://doi.org/10.1002/0470016043;CSUBTYPE:STRING:MONOGRAPH;WEBSITE:WEBSITE:PERICLES;ISPURCHASABLE:BOOLEAN:TRUE;PAGE:STRING:BOOK
9. T.G. Mayerhöfer, H. Mutschke, J. Popp, Employing theories far beyond their limits—the case of the (Boguer-) Beer-Lambert Law. Chem.Phys.Chem. **17**(13), 1948–1955 (2016). https://doi.org/10.1002/CPHC.201600114
10. T.G. Mayerhöfer, J. Popp, Beer's law derived from electromagnetic theory. Spectrochim. Acta A Mol. Biomol. Spectrosc. **215**, 345–347 (2019). https://doi.org/10.1016/j.saa.2019.02.103
11. G. Fowles, *Introduction to Modern Optics* (1989). https://books.google.com/books?hl=pt-BR&lr=&id=SL1n9TuJ5YMC&oi=fnd&pg=PA1&dq=G+R.+Fowles,+Introduction+to+Modern+Optics,+Second+Edition,+Holt,+Rinehart,+and+Winston,+New+York+(1975).&ots=DrtIluAj8K&sig=kc1k_SdFdk35xaergIfmc1UpeJk. Accessed 5 Dec 2021
12. M. Anthony, M. Fox, *Optical Properties of Solids* (Oxford University Press, 2001)
13. J. Heo, W.J. Chung, Rare-earth-doped chalcogenide glass for lasers and amplifiers, in *Chalcogenide Glasses* (2014), pp. 347–380. https://doi.org/10.1533/9780857093561.2.347
14. A.K. Somakumar, K. Upadhyay, S. Suresh, V.S. Sivasankarapillai, R. Dhanusuraman, Photoluminescent rare-earth nanocrystal-based characterization methods: advancements in photophysical applications, in *Upconversion Nanophosphors* (2022), pp. 1–18. https://doi.org/10.1016/B978-0-12-822842-5.00002-9
15. L. Giordano, D.L. Fritzen, J.M. Gonçalves, L.C.V. Rodrigues, J.M.A. Caiut, Rare-earth spectroscopy and nanomaterials, in *Modern Luminescence from Fundamental Concepts to Materials and Applications: Volume 2: Luminescence in Materials*, vol. 2 (2025), pp. 101–132. https://doi.org/10.1016/B978-0-323-88662-8.00007-3

16. A. Jha, S. Shen, M. Naftaly, Structural origin of spectral broadening of 1.5-μm emission in Er^{3+}-doped tellurite glasses. Phys. Rev. B Condens. Matter. Mater. Phys. **62**(10), 6215–6227 (2000). https://doi.org/10.1103/PHYSREVB.62.6215
17. A. Jha et al., Rare-earth ion doped TeO_2 and GeO_2 glasses as laser materials. Prog. Mater. Sci. **57**(8), 1426–1491 (2012). https://doi.org/10.1016/J.PMATSCI.2012.04.003
18. L.D. Landau, E.M. Lifshits, *Quantum Mechanics: Non-relativistic Theory* (2014), p. 688
19. S. Laroche, Exploring the quantum world: atoms, cavities, and photons. Oxford Graduate Texts **5**(11), 11 (2012)
20. V.A.G. Rivera, D. Manzani, Technological advances in tellurite glasses, in *Springer Series in Materials Science*, vol. 254 (Springer International Publishing, Cham, 2017). https://doi.org/10.1007/978-3-319-53038-3
21. J. Chacaliaza-Ricaldi, V.A.G. Rivera, I.C. Pinto, Y. Messaddeq, E. Marega, Germanate-tellurite glasses with low Er^{3+} ions concentration and their radiative emissions. J. Non Cryst. Solids **638**, 123063 (2024). https://doi.org/10.1016/j.jnoncrysol.2024.123063
22. G.C. Lozano et al., Improving laser performance at 1081 nm in Yb3+-doped tellurite glasses with low threshold. Mater. Res. Bull. **189**, 113458 (2025). https://doi.org/10.1016/j.materresbull.2025.113458
23. G. Lozano et al., Cold white light emission in tellurite-zinc glasses doped with Er^{3+}–Yb^{3+}–Tm^{3+} under 980 nm. J. Lumin. **228**, 117538 (2020). https://doi.org/10.1016/J.JLUMIN.2020.117538
24. J.L.H. Clabel et al., Overall aspects of glasses for photonic devices (2023), pp. 1–52. https://doi.org/10.1007/978-3-031-20266-7_1
25. O.L. Malta, L.D. Carlos, Intensities of 4f–4f transitions in glass materials. Quim. Nova **26**(6), 889–895 (2003). https://doi.org/10.1590/S0100-40422003000600018
26. P.L. Pernas, E. Cantelar, Emission and absorption cross-section calculation of rare earth doped materials for applications to integrated optic devices. Phys. Scr. **2005**(T118), 93–97 (2005). https://doi.org/10.1238/Physica.Topical.118a00093
27. S.O. Adeleye et al., A review of the physical, optical and photoluminescence properties of rare earth ions doped glasses. Trends Sci. **21**(12), 8759 (2024). https://doi.org/10.48048/tis.2024.8759
28. S. Tanabe, Glass and rare-earth elements: a personal perspective. Int. J. Appl. Glass Sci. **6**(4), 305–328 (2015). https://doi.org/10.1111/ijag.12142
29. M.C. Paul et al., Recent developments in rare-earths doped nano-engineered glass based optical fibers for high power fiber lasers. Trans. Indian Ceram. Soc. **75**(4), 195–208 (2016). https://doi.org/10.1080/0371750X.2016.1251852
30. S. Li, M. Zhang, Y. Peng, Q. Zhang, M. Zhao, Rate equation model analysis on the infrared and upconversion emission of Er/Yb co-doped borate-silicate glass. J. Rare Earths **28**(2), 237–242 (2010). https://doi.org/10.1016/S1002-0721(09)60087-9
31. Z. Liu, J. She, B. Peng, Spectroscopic properties of Er^{3+}-doped fluoroindate glasses. J. Rare Earths **40**(7), 1037–1042 (2022). https://doi.org/10.1016/j.jre.2021.05.011
32. D.C. Brown, N.S. Tomasello, C.L. Hancock, Absorption and emission cross-sections, Stark energy levels, and temperature dependent gain of Yb:QX phosphate glass. Opt. Express **29**(21), 33818 (2021). https://doi.org/10.1364/OE.435615
33. R.V. Barde, Influence of CeO_2 content on complex optical parameters of phosphovanadate glass system. Spectrochim. Acta A Mol. Biomol. Spectrosc. **153**, 160–164 (2016). https://doi.org/10.1016/J.SAA.2015.08.023
34. O.B. Silva, V.A.G. Rivera, Y. Ledemi, Y. Messaddeq, E. Marega, Germanium concentration effects on the visible emission properties of Er^{3+} in tellurite glasses. J. Lumin. **232**, 117808 (2021). https://doi.org/10.1016/j.jlumin.2020.117808
35. P.Y. Yu, M. Cardona, *Fundamentals of Semiconductors* (Springer Berlin Heidelberg, Berlin, Heidelberg, 2010). https://doi.org/10.1007/978-3-642-00710-1
36. J. Tauc, A. Menth, States in the gap. J. Non Cryst. Solids **8–10**(C), 569–585 (1972). https://doi.org/10.1016/0022-3093(72)90194-9
37. N.F. Mott, Electronic process in non-crystalline materials. J. Non Cryst. Solids **1**, 1 (1968)

38. Y. Wang, S. Dai, F. Chen, T. Xu, Q. Nie, Physical properties and optical band gap of new tellurite glasses within the TeO_2–Nb_2O_5–Bi_2O_3 system. Mater. Chem. Phys. **113**(1), 407–411 (2009). https://doi.org/10.1016/J.MATCHEMPHYS.2008.07.117
39. M.H.M. Zaid, K.A. Matori, S.H. Abdul Aziz, A. Zakaria, M.S.M. Ghazali, Effect of ZnO on the physical properties and optical band gap of soda lime silicate glass. Int. J. Mol. Sci. **13**(6), 7550–7558 (2012). https://doi.org/10.3390/IJMS13067550
40. M. Taj, S.C. Devaraja, U. Deka, A review of boron oxide glasses infused with individual oxides of lanthanide elements: for their physical, structural, optical, and gamma shielding properties. J. Alloys Compd. **1010**(178279) (2025). https://doi.org/10.1016/J.JALLCOM.2024.178279
41. F. El-Diasty, F.A. Abdel Wahab, M. Abdel-Baki, Optical band gap studies on lithium aluminum silicate glasses doped with Cr^{3+} ions. J. Appl. Phys. **100**(9) (2006). https://doi.org/10.1063/1.2362926/918027
42. B.D. Viezbicke, S. Patel, B.E. Davis, D.P. Birnie, Evaluation of the Tauc method for optical absorption edge determination: ZnO thin films as a model system. Physica Status Solidi (B) **252**(8), 1700–1710 (2015). https://doi.org/10.1002/pssb.201552007
43. M. Abdel-Baki, F.A. Abdel-Wahab, F. El-Diasty, One-photon band gap engineering of borate glass doped with ZnO for photonics applications, J. Appl. Phys. **111**(7) (2012). https://doi.org/10.1063/1.3698623/817025
44. M. Kranjčec, I.P. Studenyak, M.V. Kurik, On the Urbach rule in non-crystalline solids. J. Non Cryst. Solids **355**(1), 54–57 (2009). https://doi.org/10.1016/J.JNONCRYSOL.2008.03.051
45. E. Ugur, M. Ledinský, T.G. Allen, J. Holovský, A. Vlk, S. De Wolf, Life on the Urbach edge. J. Phys. Chem. Lett. **13**(33), 7702–7711 (2022). https://doi.org/10.1021/ACS.JPCLETT.2C01812/ASSET/IMAGES/LARGE/JZ2C01812_0004.JPEG
46. A.R. Zanatta, Revisiting the optical bandgap of semiconductors and the proposal of a unified methodology to its determination. Sci. Rep. **9**(1), 1–12 (2019). https://doi.org/10.1038/s41598-019-47670-y
47. I.A. Weinstein, A.F. Zatsepin, V.S. Kortov, Effects of structural disorder and Urbach's rule in binary lead silicate glasses. J. Non Cryst. Solids **279**(1), 77–87 (2001). https://doi.org/10.1016/S0022-3093(00)00396-3
48. F. Urbach, The long-wavelength edge of photographic sensitivity and of the electronic absorption of solids. Phys. Rev. **92**(5), 1324 (1953). https://doi.org/10.1103/PhysRev.92.1324
49. K.N. Dalby, H.W. Nesbitt, V.P. Zakaznova-Herzog, P.L. King, Resolution of bridging oxygen signals from O 1s spectra of silicate glasses using XPS: implications for O and Si speciation. Geochim. Cosmochim. Acta **71**(17), 4297–4313 (2007). https://doi.org/10.1016/J.GCA.2007.07.005
50. J.L. Clabel, H.G. Lozano, E. Marega, V.A.G. Rivera, XPS analysis of bridging and non-bridging oxygen in Yb^{3+}-Er^{3+}-Tm^{3+}-doped zinc-tellurite glasses. J. Non Cryst. Solids **553**, 120520 (2021). https://doi.org/10.1016/J.JNONCRYSOL.2020.120520

Chapter 3
Emission Spectrum

Chapter 2 discusses that a REI reaches an exciting state after absorbing a certain energy. The phenomenon in which the REI undergoes a radioactive transition releasing energy in the form of light is called luminescence.[1] This is a relevant phenomenon for optical technologies that use glasses, e.g., optical communications [2], solid-state lightning [3], lasers [4], random lasers [5], optical amplifiers [6], and sensors [7]. Therefore, understanding the mechanisms behind the emission process of REI is paramount to the development of suitable REI-doped glasses for such applications. In this framework, no excitation wavelength produces the emission of a REI, and this is one of the main reasons to collect the absorption spectrum and identify the absorption bands. However, excitement with these different absorption wavelengths will give different responses in the emission spectra. Concerning this point, an excitation spectroscopy is required to recognize the excitation wavelengths for a specific emission band to obtain an efficient luminescence performance.

The REI exhibits severe energy levels and offers a wide variety of emission wavelengths from ultraviolet to middle infrared. For instance, a common Er^{3+}: $^4I_{13/2} \rightarrow {}^4I_{15/2}$ emission centered at $\approx$1530 nm is used in optical communications, e.g., erbium-doped fiber amplifier (EDFA) [8], Yb^{3+}: $^2F_{5/2} \rightarrow {}^2F_{7/2}$ centered at $\approx$1000 nm for solid-state lasers [9], Nd^{3+}: $^4F_{3/2} \rightarrow {}^4I_{11/2}$ emission centered at 1060 nm for optical amplifiers and gain media [10], and Tm^{3+}: $^3F_4 \rightarrow {}^3H_6$ emission centered at $\approx$1810 nm [11]. In consequence, REI-doped glasses are functional materials for lasing applications covering a wide spectral range. Besides, to obtain these and other emission wavelengths in REI, two luminescence processes are involved, i.e., Stokes and anti-Stokes. Stokes luminescence refers to the emission process in which the energy of the absorbed photons is larger than the energy of the emitted photons and involves energy loss due to non-radiative (NR) decays. In anti-Stokes luminescence, higher energy

[1] In general, the excitation mechanism can be via light, thermal, chemical, electrical, mechanical, among others. When emission occurs under excitation with UV or visible light, it is called photoluminescence [1]. Since this book employs different kind of excitation light (UV, visible, and NIR), the term "luminescence" will be employed.

G. Lozano C. et al., *A Brief Guide to Optical Spectroscopy of Trivalent Rare Earth Ions in Glasses*, Springer Series in Optical Sciences 247,
https://doi.org/10.1007/978-3-032-18501-3_3

photons are emitted after absorbing lower energy photons, where the utilization of additional energy is involved, e.g., heating or another photon excitation.

The most important spectroscopic techniques for REI emission are presented with a brief description of the employed equipment, besides applications in solid-state lighting (colored and white light generation), and a detailed analysis of the REI luminescence. Through this comprehensive approach, we establish the fundamental principles governing REI emission while highlighting their practical implementation in cutting-edge photonic technologies.

3.1 Measurement of Emission Spectrum

The emission spectrum is collected with the equipment named the spectrofluorometer. Basically, it consists of an excitation source, collimators, monochromators, and a detector, which can be seen in a simplified schematics in Fig. 3.1. Herein, the excitation source, e.g., Xe lamp, which is composed of a broadband of wavelengths, illuminates a collimator lens to focus the light on a diffraction grating. In this manner, light from the source is dispersed into different wavelengths, and a slit works as an individual wavelength selector. This diffraction grating and the slit are the excitation monochromator. Once the excitation wavelength is fixed, the sample is irradiated, and emission occurs. The emitted light is then collimated and passes through an emission monochromator, which selects the emission wavelengths and intensities to be analyzed by a detector. The spectrofluorometer operates in two modes: emission and excitation spectroscopy. For emission measurements, a single excitation wavelength λ_{exc} is selected to record the emission spectrum across a specific spectral range. Conversely, for excitation spectroscopy, a fixed emission wavelength λ_{em} is monitored to identify the specific excitation wavelengths that generate that emission. According to this schematic, both diffraction gratings can rotate. In the case of emission spectroscopy, the excitation monochromator is kept fixed, while the emission monochromator rotates. In contrast, for excitation spectroscopy, the excitation monochromator rotates while the emission monochromator remains fixed.

3.2 Down-Shifting

Down-shifting (DS) is a Stokes luminescence process in which a REI reaches an exciting level after absorbing a photon and emits a photon with lower energy than the absorbed one. When a REI absorbs one photon and emits two or more photons with low energy it is called quantum cutting [12], e.g., a Pr^{3+}-Yb^{3+}-doped glass after absorbing one visible photon at 443 nm emits NIR photons at 1.0 and 2.0 μm [13]. Down-conversion is the process where a REI absorbs a high-energy photon and emits a lower energy photon [12], e.g., an Nd^{3+}-doped glass emits 448 nm upon 338 nm excitation [14]. Nevertheless, some authors consider down-conversion and

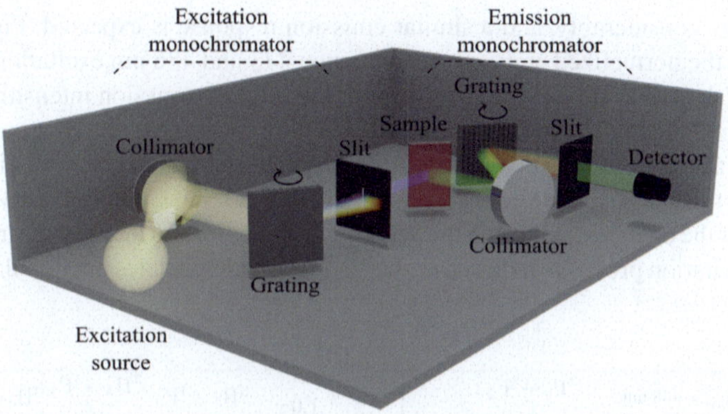

Fig. 3.1 Schematic diagram of a spectrofluorometer, where the main components are illustrated

quantum cutting as the same luminescence process [15, 16], while others consider that quantum cutting is a consequence of the cooperative down-conversion effect [17, 18]. Since both effects are particular cases, this textbook will strictly employ the term DS.

3.2.1 *Analysis of Down-Shifting and Excitation Spectra*

For a complete study of the excitation and emission processes, a tellurite glass (composed of TeO_2, ZnO, and La_2O_3) doped with 1.0 mol% of Pr^{3+} was fabricated. The Pr^{3+} was chosen due to its rich energy level diagram, which offers visible emissions upon UV excitation. A typical absorption band of the Pr^{3+} is the $^3H_4 \rightarrow {}^3P_2$ transition centered at ≈448 nm, which in tellurite glasses exhibit a relative intense absorption peak [19, 20]. Hence, this excitation wavelength will be employed to observe the Pr^{3+} emission spectrum in the visible region, which is illustrated in Fig. 3.2a. Here, the bands centered at 530, 543, 614, 647, 684, and 707 nm are ascribed to the $^3P_1 \rightarrow {}^3H_5$, $^3P_0 \rightarrow {}^3H_5$, $^3P_0 \rightarrow {}^3H_6$, $^3P_0 \rightarrow {}^3F_2$, $^3P_1 \rightarrow {}^3F_3$, and $^3P_1 \rightarrow {}^3F_4$ transitions, respectively [19]. However, 448 nm is not the only excitation wavelength to obtain the observed emission bands. In this manner, the excitation spectrum by monitoring the most representative emissions, i.e., the bands centered at 614 and 647 nm were measured and are displayed in Fig. 3.2b. Herein, three peaks centered at 448, 473, and 486 nm correspond to the $^3H_4 \rightarrow {}^3P_2$, $^3H_4 \rightarrow {}^3P_1$, and $^3H_4 \rightarrow {}^3P_0$ absorption bands. The peaks of the excitation spectra indicate the position of the wavelengths necessary to obtain a specific emission wavelength, and their relative intensities indicate how strong the emission intensity is when the REI is excited at a specific excitation wavelength. In other words, the higher the peak intensity in the excitation spectrum, the higher the efficiency to obtain an emission band. In addition, the normalized emission spectra indicates that the relative intensities of the peaks do not vary considerably, and a similar emission response is expected. Figure 3.2c illustrates the normalized emission spectra upon 473 and 486 nm excitation, and the centers of the peaks do not change, likewise the relative emission intensities do not differ independently of the excitation wavelength.

In addition to these results, the absorption spectrum, the energy level diagram of the Pr^{3+} in this glass can be designed, as seen in Fig. 3.2d. First, it is necessary to set the position of the energy states using the absorption spectrum and identify the transitions for the excitation process. In this case, the $^3P_2$, $^3P_1$, and $^3P_0$ states are populated from

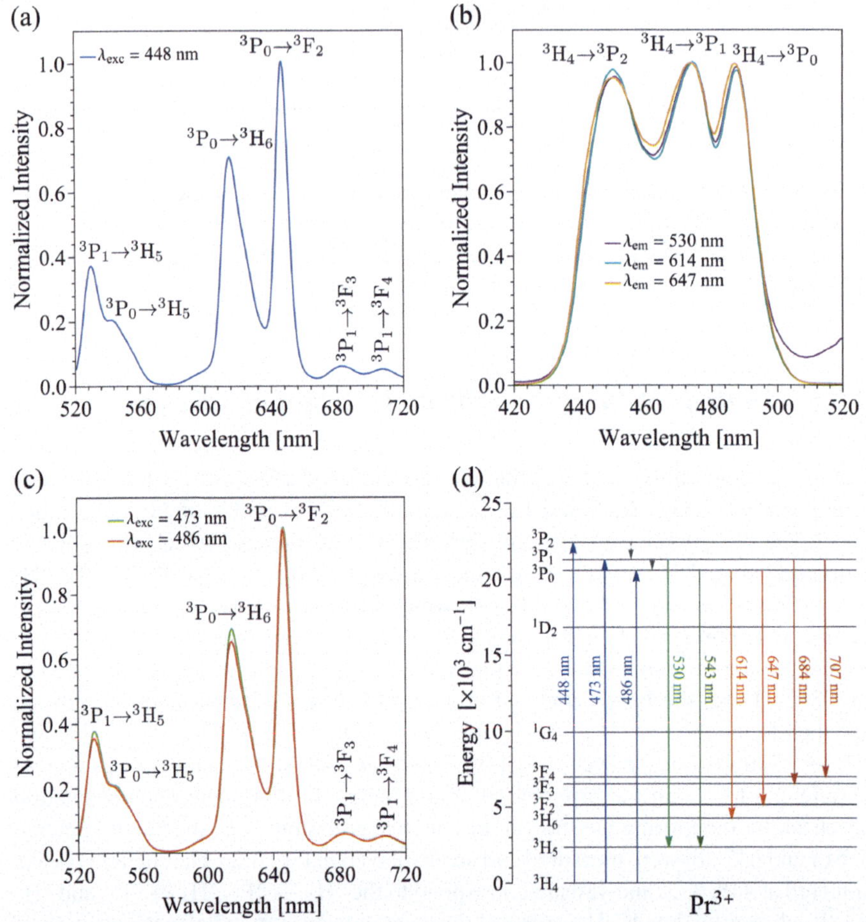

Fig. 3.2 **a** Normalized emission spectrum of the Pr^{3+} under 448 nm excitation. **b** Normalized excitation spectra monitoring the 614 and 647 nm emission bands, i.e., $^3P_0 \rightarrow {}^3H_6$ and $^3P_0 \rightarrow {}^3F_2$ transitions, respectively. **c** Normalized mission spectra of the Pr^{3+} upon 473 and 486 nm excitation. These wavelengths were obtained from the excitation spectra. **d** Energy level diagram of the Pr^{3+}, showing the excitation wavelengths and visible emissions, where the dashed downward arrows indicate non-radiative transitions. The position of the energy levels was obtained from Ref. [21]

the ground state $^3H_4$, and this process is known as ground state absorption (GSA). For instance, from the $^3P_2$ level, the $^3P_1$ and $^3P_0$ levels are populated through NR decays (phonon relaxation) giving rise to the transitions observed in the emission spectra. These transitions can be found in previous research and corroborated with the positions of the peaks in energy units of the emission spectrum. Upon 486 nm excitation, the $^3P_0$ level is populated, but transitions from the next excited state $^3P_1$ are observed (anti-Stokes luminescence). The last is because the small energy gap between the $^3P_1$ and $^3P_0$ levels, which according to the excitation spectra is approximately 566 cm^{-1}, and the $^3P_1$ state is thermally populated by processes supported by phonons [22]. The design of an energy level diagram is a powerful tool to visualize and to explain the transition pathways of the excitation and de-excitation dynamics of REI in glasses. In this section, it was proposed that a simplified energy level diagram should be used; however, as the next chapters unfold, more complicated processes will be explained, and the transition pathways will become more accurate involving all the spectroscopic processes.

3.3 Up-Conversion

Up-conversion (UC) is an anti-Stokes and non-linear luminescence process where an atom, ion, or molecule absorbs two or more photons with a specific frequency and stimulates the emission of a photon with a frequency larger than the initial photons. This phenomenon was first experimentally observed in Er^{3+}–Yb^{3+}-doped sodium tungstate glasses by François Auzel in 1966 [23]. In the UC process, the absorption of the photons must be sequential and not simultaneous, in contrast to two-photon absorption and second/third/fourth... harmonic generation.[2] Besides, different UC processes are identified and are summarized in Fig. 3.3, i.e., excited-state absorption (ESA), energy transfer UC (ETU), cooperative sensitization (CS), cooperative luminescence (CL), cross-relaxation (CR), and photon avalanche (PA), and all these processes shall be briefly described considering two-photon UC in a three-level system.

The ESA process occurs in one REI and consists of successive absorption of photons. Here, a REI is first excited until a metastable energy level via GSA, and then the next upper level is populated. Finally, the UC emission is obtained when the last exciting level decays to the ground state. ETU, which was first named by Auzel as APTE effect (French acronym for *addition de photon par transferts d'energie*), is an energy transfer (ET) process from one REI to another one, where the involved photons must be in resonance in each excitation step. ETU can occur from one REI to another who is already in an exciting state or from two REI as seen in Fig. 3.3. Here, the ETU in REI is the most efficient UC process since photons have to interact with the medium a longer time, which is achieved under resonances [26]. CS is another

[2] Since the second-harmonic generation is observed in REI-doped glasses [24, 25], this process will not be covered in the textbook.

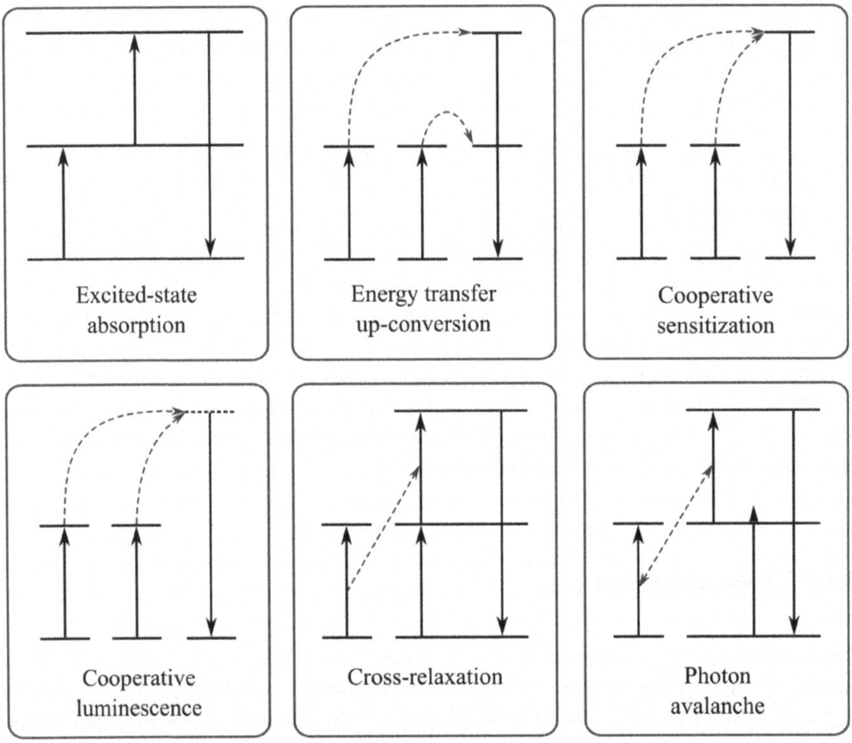

Fig. 3.3 Energy level diagram describing the different up-conversion processes. Curved dashed lines indicate ET process and straight dashed lines indicate CR process

kind of ET process, in which, in this case, two excited REIs transfer their energy to another REI. In the case of CL, only two REIs are involved since each one absorbs one photon and cooperates to generate the emission. CR occurs when an excited REI transfers its energy to another excited REI and via ESA reaches the third energy level. CR can occur between identical REI or between REIs with a similar energy gap between their energy levels. The probability of observing CR is greater when the doping concentration of REI is higher, and it is associated with the interaction distance between REIs [27]. PA is a relatively complex process since it involves ET via repetitive CR, and a non-resonant GSA (that is why the upper arrow is slightly above the intermediate energy level of the REI from the right). Besides, this effect is observed in the low excitation power regime and with large concentrations of REI [28].

Furthermore, it is possible to experimentally determine the number of photons involved in the UC process: There is a relationship between the UC emission intensity, denoted as I_{UC}, and the excitation intensity I_{exc}, i.e., $I_{UC} \propto I_{\text{exc}}^n$, where n is the number of photons involved in the population inversion of the emission band [29]. In this way, it is reflected in the non-linear behavior of the UC process in a low-power regime. To calculate the number of pumping photons, it is necessary to collect the UC spectra at different excitation powers/intensities and calculate the maximum intensity or the area of each UC band. Depending on the author, it is used in the area or the maximum intensity of the bands; however, it is important to remark that most of the REI emission bands exhibit more than one peak and a bandwidth generated by the Stark splitting. The calculation of the band area implies the contribution of all Stark levels of the analyzed emission band. In this section, the number of NIR photons involved in each UC emission band using both maximum intensity and integrated band area will be covered.

3.3.1 *Analysis of the Up-Conversion Spectrum*

Here, a tellurite glass (composed of TeO_2, ZnO, K_2O, and GeO_2) is doped with 1.0 mol% of Er^{3+} which was employed for these calculations. The sample was subjected to an excitation of 980 nm using a commercial diode laser with tunable excitation power. This excitation wavelength is commonly employed because the Er^{3+} exhibits an absorption band centered at ≈976 nm ($^4I_{15/2} \rightarrow {}^4I_{11/2}$), and with the energy gap between other intermediate levels (≈10,200 cm^{-1}) [30]. Therefore, the characteristic visible emissions of the Er^{3+}, i.e., green and red light, can be obtained via UC.

The sample was pumped at different excitation powers P_{exc} from 20 to 100 mW, as displayed in Fig. 3.4a. Herein, the emission bands centered at 527, 543, and 657 nm are ascribed to the $^2H_{11/2} \rightarrow {}^4I_{15/2}$, $^4S_{3/2} \rightarrow {}^4I_{15/2}$, and $^4F_{9/2} \rightarrow {}^4I_{15/2}$ transitions. It can be observed that the emission intensities of the band centered at 543 nm are slightly larger than the 657 nm emission band for all excitation powers. Besides, no significant variations in the intensities of each emission band are observed varying the pumping power. This last should indicate that the UC performance does not exhibit luminescence quenching in this excitation power region. Further, according to the spectrum from Fig. 3.4b, the $^2H_{11/2} \rightarrow {}^4I_{15/2}$ and $^4S_{3/2} \rightarrow {}^4I_{15/2}$ bands are overlapped, and a trough is centered at ≈535 nm (see the inset from Fig. 3.4a). This point is important because both bands are used for the integrated area calculations. In this way, the three areas are denoted as A1, A2, and A3, and their numerical values will be computed by direct integration. Once the maximum intensities of the peak wavelengths (I_{max}) and the band area are determined, a plot of these quantities as a function of the excitation power is constructed, as shown in Fig. 3.4c. For simplicity, the logarithm of the I_{max}, the areas and the P_{exc} were evaluated to perform a linear fitting and calculate the slopes, i.e., n. As mentioned before, the maximum intensities of the bands centered at 543 and 657 nm are moderately similar, and their values are nearly overlapped. Regarding the band area, there is a considerable difference between the 543 and 657 nm bands. This is because the 657 nm emission band presents a larger bandwidth. In addition, for both, I_{max} and areas, the slopes obtained by linear fitting for all bands ($R^2 > 0.99$) can be approximated to 2, since n must be an integer, and these results are in agreement with other Er^{3+}-doped tellurite glasses [31–35]. Therefore, same results are achieved considering our maximum peak intensities or band areas. Nevertheless, some experimental considerations have to be highlighted: electrical noise, integration time, wavelength step of the monochromator, and heating effects may alter the peak position of the bands. Moreover, each band displays Stark levels, which their relative intensities depend on the light source and the glass matrix and then may be a source of inaccuracy on the peak position calculation. For these reasons, the authors consider that the calculation of the integrated area is the most suitable method to determine the number of NIR photons involved in the UC process. For more details about the advantages of calculating the band area, the reader is referred to in Chap. 5.

Figure 3.4d illustrates a simplified energy level diagram which summarizes the population of the different energy states via UC process, where two 980 nm photons are involved in the UC emission transitions. For the green emissions, the possible mechanism is the GSA to the $^4I_{13/5}$ state and then an ESA to the $^4F_{7/2}$ state. From this level, consecutive NR transitions populate the $^2H_{11/2}$ and $^4S_{3/2}$ states, and from these states, the 527 and 543 nm emissions are obtained after decay to the $^4I_{15/2}$ ground state. Additionally, the red emission can be obtained after successive NR transmissions from the $^4F_{7/2}$ state to the $^4F_{9/2}$ state. Another way to populate the $^4F_{9/2}$ level is a GSA to the $^4I_{11/2}$, where an NR decay to the $^4I_{13/2}$ level occurs, and via ESA with another 980 nm photon the $^4F_{9/2}$ level is populated.

3.4 Emission Cross-Section

Overall, likely to the absorption cross-section, the emission cross-section (σ_{em}) quantifies the probability or rate of release of electromagnetic radiation of a material after the excitation process. Both quantities are valuable for evaluating the lasing performance of optical gain mediums, where larger values are required for optical applications. Different models were developed to calculate σ_{em} for REI, e.g., McCumber relation, Füchtbauer–Ladenburg relation, and direct measurements of small signals or saturated gain [36]. This guide is focused on the McCumber and Füchtbauer–Ladenburg relations since they are the most employed for REI optical spectroscopy.

3.4.1 McCumber and Füchtbauer–Ladenburg Relations

The emission cross-section of the 1.53 μm emission band of the Er^{3+} is crucial to evaluate the suitability of EDFA. Therefore, for the calculations in this section, the $^4I_{13/2} \rightarrow {}^4I_{15/2}$ emission of the Er^{3+}-doped glass at room temperature shall be used.

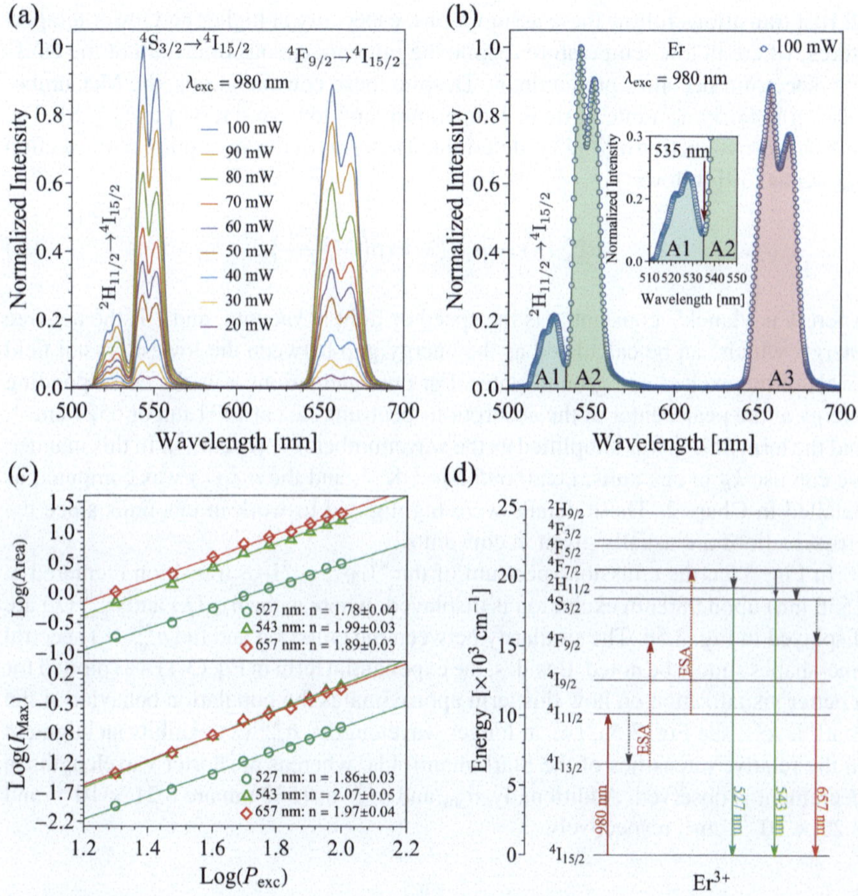

Fig. 3.4 **a** Normalized UC spectra of a Er^{3+}-doped tellurite glass upon 980 nm excitation at different excitation powers. **b** Normalized UC spectrum of a Er^{3+}-doped tellurite glass upon 980 nm excitation at a pumping power of 100 mW, showing the areas of each emission band (A1, A2, and A3). *Inset*: Zoom of the spectrum for a better visualization of the trough centered at 535 nm (indicated by an arrow). **c** Dependence of the logarithm of the maximum intensity and the integrated area and of each band with respect to the excitation power P_{exc}. Legends indicate the emission band wavelength, the slope values, and their fitting error. **d** A simplified energy level diagram of the Er^{3+}-doped glass, where the dashed arrows indicate non-radiative transitions

3.4.1.1 McCumber Relation

The McCumber relation, known as reciprocity relation, involves thermodynamic principles and the Einstein relations of absorption, spontaneous and stimulated emission. This relation is widely employed in systems with narrow and manifolds spectra due to the Stark effect, and with homogeneous linewidth. In addition, it is assumed that the energy width of each Stark level is lower than the thermal energy $k_B T$ [37], where k_B is Boltzmann's constant and T is the absolute temperature. However, not all REI transitions follow these assumptions, especially at higher and lower temperatures, where in low-temperature regime the inhomogeneous linewidth of the emission spectrum becomes predominant. Despite these considerations, the McCumber relation is fairly accurate at room temperature and low excitation power [38]. The emission cross-section $\sigma_{em}^{MC}(\lambda)$ is determined in terms of the absorption cross-section $\sigma_{abs}(\lambda)$ as [39] follows:

$$\sigma_{em}^{MC}(\lambda) = \sigma_{abs}(\lambda)\exp\left(\frac{\epsilon - hc\lambda^{-1}}{k_B T}\right) \tag{3.1}$$

where h is Planck's constant, c is the speed of light at vacuum, and ϵ is the net-free energy, which can be calculated as the energy gap between the lowest crystal field levels of the excited and ground states. For our calculations, ϵ is the corresponding energy at the peak center of the absorption spectrum, i.e., at 1531 nm or 6527 cm^{-1}, and the term $hc\lambda^{-1}$ was simplified as the wavenumber λ^{-1} (in cm^{-1}). In this manner, we can use k_B in cgs units, i.e., 0.695 cm^{-1} K^{-1}, and the $\sigma_{abs}(\lambda)$ was computed as detailed in Chap. 2. These details were highlighted to work in cgs units since the cross-sections are usually given in cm^2 units.

In Fig. 3.5a, the emission spectrum of the $^4I_{13/2} \rightarrow {}^4I_{15/2}$ transition (centered at 1536 nm) upon 980 nm excitation is displayed, whereas the $\sigma_{abs}(\lambda)$ and $\sigma_{em}^{MC}(\lambda)$ are displayed in Fig. 3.5b. The similarity between the emission and the $\sigma_{em}^{MC}(\lambda)$ spectral line-shapes should be noted. Besides, the exponential term in Eq. (3.1) was plotted for a better visualization on how this term approximates the population behavior of the Stark levels, see Fig. 3.5a, i.e., at longer wavelengths, $\sigma_{em}^{MC}(\lambda)$ exhibits an augment in the relative intensities of the Stark manifolds, whereas at shorter wavelengths, a decrement is observed. Additionally, σ_{abs} and σ_{em}^{MC} at 1532 nm are 8.21×10^{-21} and 8.25×10^{-21} cm^2, respectively.

3.4.1.2 Füchtbauer–Ladenburg Relation

The Füchtbauer–Ladenburg relation describes the connection between the emission cross-section σ_{em}^{FL} and the emission spectrum, which was determined considering the Einstein A and B coefficients. This relation is expressed as follows [38]:

$$\sigma_{em}^{FL}(\lambda) = \frac{\lambda^4}{8\pi c n^2 \tau}\frac{I(\lambda)}{\int I(\lambda)d\lambda} \tag{3.2}$$

where n is the refractive index, τ is the radiative lifetime, and $I(\lambda)$ is the normalized line-shape of the emission spectrum. Since Eq. (3.2) requires the experimental luminescence spectrum, the lifetime, and the refractive index of the glasses, measurement errors interfere with the accuracy of the σ_{em}^{FL} calculation. Furthermore, semi-empirical calculations can be performed to estimate the inverse of the lifetime, i.e., the radiative transition rate or transition probability, which is computed using the Judd–Ofelt

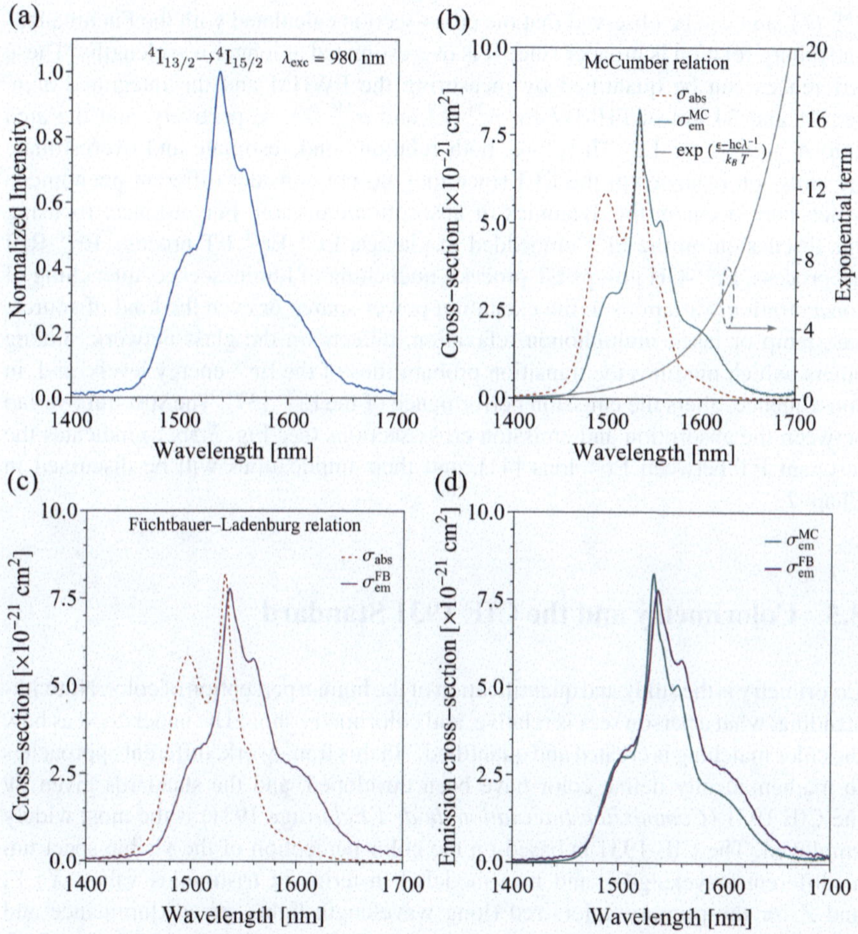

Fig. 3.5 **a** Emission spectrum of the $^4I_{13/2} \rightarrow {}^4I_{15/2}$ transition under 980 nm excitation. **b** Absorption and emission cross-section (using McCumber relation) of the $^4I_{13/2} \rightarrow {}^4I_{15/2}$ band, and the exponential term from Eq. (3.1). **c** Absorption and emission cross-section (using the Füchtbauer–Ladenburg relation) of the $^4I_{13/2} \rightarrow {}^4I_{15/2}$ band. **d** Comparison between the emission cross-sections calculated from McCumber and Füchtbauer–Ladenburg relations

theory (Chap. 9) [40]. Hence, a comparison between the McCumber and the Füchtbauer–Ladenburg relations should be performed to qualitatively analyze the lasing performance of the glass or other optical gain medium.

Both, $\sigma_{abs}(\lambda)$ and $\sigma_{em}^{FL}(\lambda)$, are displayed in Fig. 3.5c, and a slight wavelength shift is distinguished because the emission of light involves vibrational transitions between the Stark manifolds of the $^4I_{13/2}$ energy level, leading to changes in the peak position compared to the absorption peaks. Herein, the σ_{em}^{FL} at 1536 nm is 7.79×10^{-21} cm^2, lower than reported with the McCumber relation. Figure 3.5d shows the $\sigma_{em}^{FL}(\lambda)$ and

$\sigma_{em}^{MC}(\lambda)$, and can be observed that the cross-section calculated with the Füchtbauer–Ladenburg relation is broader since it is overestimated at longer wavelengths. These differences can be quantified by measuring the FWHM and the integrated area, i.e., 41 and 54 nm of FHMW for $\sigma_{em}^{FL}(\lambda)$ and $\sigma_{em}^{MC}(\lambda)$, respectively, and the area ratio $A_{\sigma_{em}^{FL}}/A_{\sigma_{em}^{MC}} = 1.2$. Therefore, both relations underestimate and overestimate the emission behavior of the REI since they do not consider different phenomena which may occur in the dynamics of absorption/emission. For instance, focusing this discussion on the Er^{3+} embedded in glasses, Er^{3+}-Er^{3+} ET process, Er^{3+}-REI ET process, Er^{3+}-OH groups ET process, quenching of luminescence, quenching of concentration, variations in the excitation power source or even the kind of source, e.g., lamp or laser, multiphonon relaxation, defects on the glass network, among others, which modifies the transition probabilities of the Er^{3+} energy levels, and, in consequence, alters the emission performance of the Er^{3+} [37]. The spectral overlap between the absorption and emission cross-sections (see Fig. 3.5b, c) indicates the resonant ET between Er^{3+} ions [41], and their implications will be discussed in Chap. 7.

3.5 Colorimetry and the CIE 1931 Standard

Colorimetry is the study and quantification of the human perception of color. Notwithstanding, what a person sees is relative, and colorimetry should be understood as how the color matching is created and quantified.[3] In this framework, different approaches to mathematically define color have been developed, and the standards given by the CIE 1931 (*Commission Internationale de l'Eclairage* 1931) is the most widely employed. The CIE 1931 is based on the color perception of the visible spectrum at different wavelengths, and it is modeled in terms of tristimulus values X, Y, and Z for the primary colors red (long wavelength light), green (luminance and mid-wavelengths), and blue (short wavelengths light), or RGB, respectively, which represent the CIE XYZ color space.[4] Since all colors we see are a mixture of the RGB, the color perception is defined by tristimulus curves using the color matching functions $\overline{x}$, $\overline{y}$, and $\overline{z}$, and are plotted in Fig. 3.6a. These functions were experimentally determined by averaging the responses of observers to monochromatic light stimuli.

The CIE XYZ color space can be further simplified into a 2D chromaticity diagram by normalizing the tristimulus values with the color matching functions as $x = \frac{\overline{x}}{\overline{x}+\overline{y}+\overline{z}}$ and $y = \frac{\overline{y}}{\overline{x}+\overline{y}+\overline{z}}$ (see Fig. 3.6b). This diagram maps all perceivable colors, with its curved boundary, i.e., the spectrum locus, representing the pure monochromatic

[3] The authors highly recommend Ref. [42] for a further discussion about this topic.

[4] Initially, the CIE 1931 RGB color space was developed, however, red light was constructed using green and blue light, according to the perception of different observers. This led to negative values in the model of the red color spectrum. To address this issue, the CIE 1931 XYZ color space was eventually formulated [43].

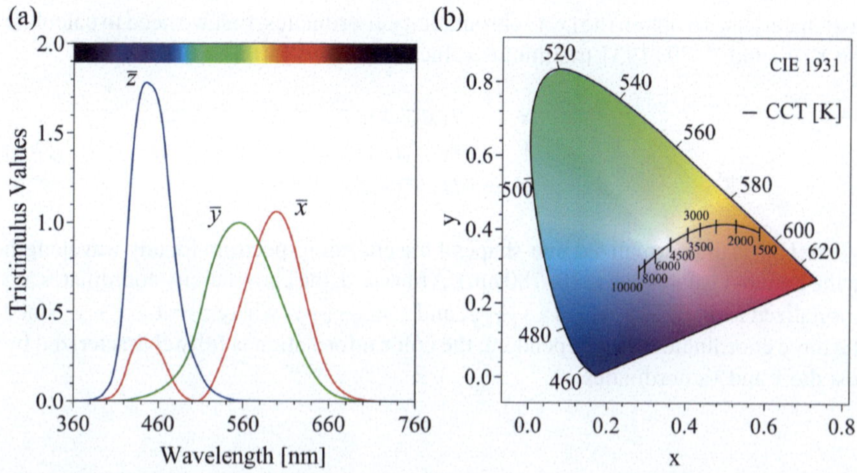

Fig. 3.6 **a** Color matching functions plot, represented as $\bar{x}$, $\bar{y}$, and $\bar{z}$.. The colored rectangle at the top of the plot indicates the RGB colors as a function of wavelength according to the CIE 1931 standards. **b** CIE 1931 chromaticity diagram of the RGB primaries, showing the CCT curve (in K). The color white is at (0.333, 0.333). The spectrum locus wavelengths are located at the boundaries of the color space

colors of the visible, each point corresponding to a specific wavelength. Further, in Fig. 3.6b is displayed a curve connecting the correlated color temperatures (CCT of blackbody radiators (in K)) [44]. This curve is useful to represent artificial light sources, e.g., fluorescent lamps and light-emitting diodes (LEDs), even though their spectral power distribution being different from a blackbody source. The CCT is employed to describe white light sources, and we will discuss this topic in detail later. Notably, the equal-energy white point (E) is represented in the CIE 1931 chromaticity diagram at coordinates (0.333, 0.333).

3.5.1 Calculating the CIE (x, y) Coordinates

After an appropriate excitation, several REIs emit at different wavelengths, with most emissions occurring in the visible region. For instance, we observe intense green emission in Er^{3+}-doped tellurite and phosphate glasses [45, 46], red emission in Eu^{3+}-doped silicophosphate glasses [47], blue emission in Tm^{3+}-doped phosphate glasses [48], or orange/yellow emissions in Pr^{3+}-doped borate glasses [49]. Each REI of the mentioned examples exhibits distinct emission spectral profiles. These differences manifest in both relative emission intensities and spectral line-shapes, which can be mathematically transformed into (x, y) chromaticity coordinates in the CIE 1931 color space. This conversion provides an objective quantification of their color emission properties, enabling comparison between different REI dopants and host materials. To obtain the (x, y) chromaticity coordinates, first we need to calculate the X, Y, and Z CIE 1931 tristimulus values as follows [50]:

$$\begin{aligned} X &= \int I(\lambda)\bar{x}(\lambda)d\lambda \\ Y &= \int I(\lambda)\bar{y}(\lambda)d\lambda \\ Z &= \int I(\lambda)\bar{z}(\lambda)d\lambda \end{aligned} \tag{3.3}$$

where $I(\lambda)$ is the normalized line-shape of the emission spectrum for any wavelength in the human visual range (380–780 nm). Afterward, the chromaticity coordinates are normalized as $x = \frac{X}{X+Y+Z}$, $y = \frac{Y}{X+Y+Z}$, and $z = \frac{Z}{X+Y+Z}$, where $z = 1 - x - y$. Since the three coordinates are independent, the color information is fully characterized by just the x and y coordinates.

3.5.2 *Colored Light Emission*

For this section, some zinc-tellurite glasses all of them doped with Yb^{3+} (1.0 mol%), and with different combinations of Er^{3+} (0.4 mol%), Tm^{3+} (0.8 mol%), and Ho^{3+} (0.3 mol%), i.e., Er^{3+}, Tm^{3+}, Ho^{3+}, Er^{3+}-Tm^{3+}, Ho^{3+}-Er^{3+}, Er^{3+}-Ho^{3+}, Ho^{3+}-Tm^{3+}, and Er^{3+}-Ho^{3+}-Tm^{3+}. The samples are labeled according to the REI combinations, and the Yb^{3+} is employed as sensitizer upon 980 nm excitation for UC emissions at a given excitation power. All normalized spectra are depicted in Fig. 3.7 and are separated by single, bi, and triply doped samples in Fig. 3.7a–c, respectively. From Fig. 3.7a, we identify in the Tm sample the bands centered at 474, 649, and 690 nm, ascribed to the Tm^{3+}: ${}^1G_4 \rightarrow {}^3H_6$, ${}^5F_4 \rightarrow {}^3H_6$, and ${}^3F_3 \rightarrow {}^3H_6$ transitions, respectively. For the Er sample, the bands centered at 521, 543, and 655 correspond to the Er^{3+}: ${}^2H_{11/2} \rightarrow {}^4I_{15/2}$, ${}^4S_{3/2} \rightarrow {}^4I_{15/2}$, and ${}^4F_{9/2} \rightarrow {}^4I_{15/2}$ transitions, respectively, whereas for the Ho sample, the bands at 542 and 658 nm are attributed to the Ho^{3+}: ${}^5F_4 \rightarrow {}^5I_8$ and ${}^5F_5 \rightarrow {}^5I_8$ transitions, respectively. All spectra are normalized because, according to Eq. (3.3), the tristimulus values depend on the line-shape of the spectrum. In addition, all measurements were performed in the wavelength region from 450 to 750 for two main reasons: First, before 450 nm there are no emission peaks via UC in all samples. Second, if we observe Fig. 3.7a, above 750 nm the color matching functions are zero since beyond 700 nm they fall within the NIR region. Therefore, they are not included in the calculations.

In Fig. 3.7b, we can see the spectra of the bi-doped samples, where changes in the relative intensities bands are noted. This is because there is ET process between the REIs, which promotes the population and depopulation of the emission bands involved. Here, a notorious example is depicted in the ErTm and HoTm samples, where the blue emission band of the Tm^{3+} ${}^1G_4 \rightarrow {}^3H_6$ is drastically reduced. In all bi-doped samples, the red emission is favored because Er^{3+}, Tm^{3+}, and Ho^{3+} display red emission bands. It should be noted that the relative intensities of these emissions can be tuned by adjusting the excitation power and varying the REI doping

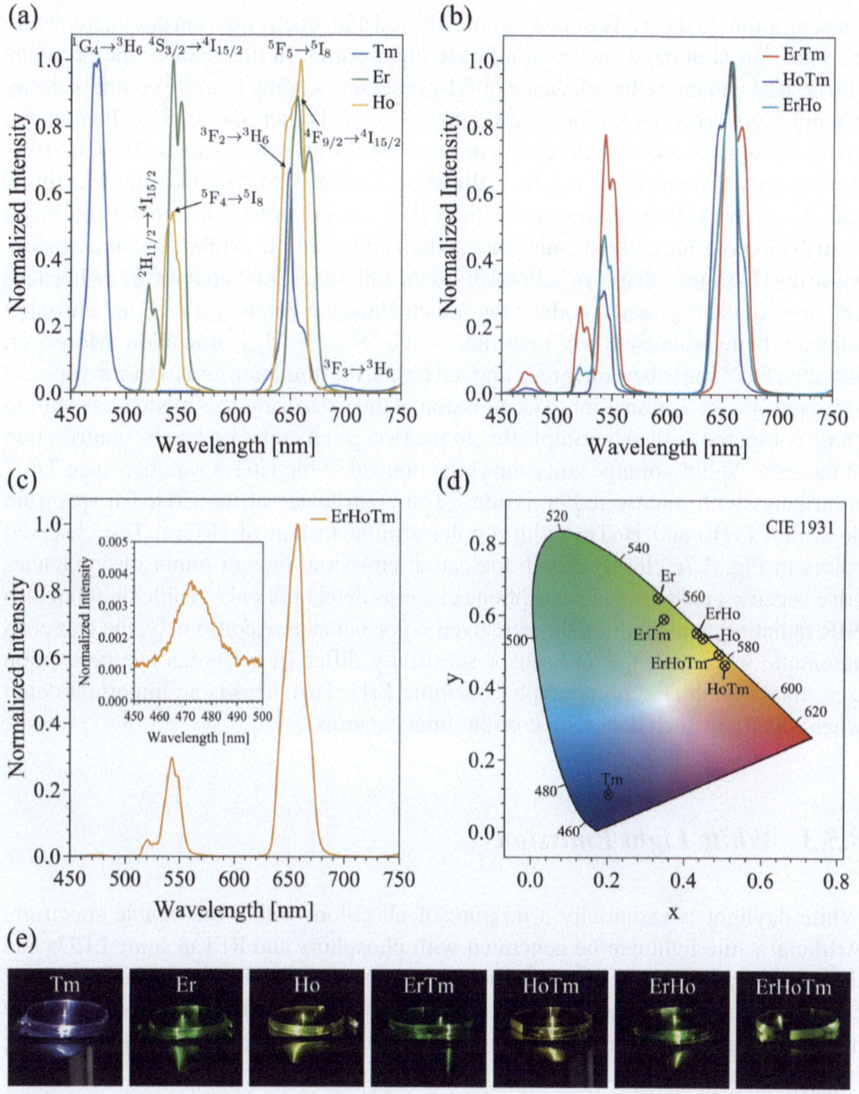

Fig. 3.7 UC emission spectra of the **a** Tm, Er, Ho, **b** ErTm, HoTm, ErHo, and **c** ErHoTm samples. *Inset*: Magnification of the spectral region, corresponding to the Tm^{3+}: $^1G_4 \rightarrow {}^3H_6$ emission band. **d** CIE 1931 chromaticity diagram displaying the calculated coordinates (*x*, *y*) of the UC spectra. **e** Photograph of the samples studied captured using a conventional smartphone camera

concentration. In the triply doped sample, the red and green emission dominate, while the blue emission band shows significant suppression. In this system, the presence of four REI enhances the relevance of ET processes, leading to more complex decay dynamics governed by ET, population inversion, and other interrelated phenomena. In this section, we focus on the color emission properties of the glasses. The CIE 1931 chromaticity diagram of all spectra is illustrated in Fig. 3.7d. The Tm sample exhibits emission near the blue region, and a slight shift toward violet is due to the combined contribution of blue and red emission bands. The Er sample exhibits green emission, while the Ho sample displays yellowish-green emission. This behavior arises because Er^{3+} ions in these glasses produce two green emission bands, with the red emission intensity being slightly lower than that of the $^4S_{3/2} \rightarrow {}^4I_{15/2}$ transition. Moreover, although Ho^{3+} ions also emit green and red lights, the dominant emission corresponds to the red $^5F_5 \rightarrow {}^5I_8$ transition. The emission of the ErHo sample is moderately shift to green compared to the Ho sample due to the two green emission bands' contribution of the Er^{3+}. Yellow-orange emissions are obtained in the HoTm sample, since Tm^{3+} contributes with intense red emissions. The coordinates of the ErHoTm spectrum lie within ErHo and HoTm, with a color similar to that of HoTm. The observed colors in Fig. 3.7e closely match the actual emission, though minor discrepancies arise because conventional smartphone cameras detect not only visible light but also NIR radiation, which alters the perceived color balance. Additionally, the camera's automatic white balance and sensor sensitivity differ from human eye perception (see, for example, the photograph of sample ErHoTm). This is an important detail when recording such data with conventional cameras.

3.5.3 White Light Emission

White daylight is essentially a mixture of all colors within the visible spectrum. Artificial white light can be generated with phosphors and REI in some LEDs and lamps. For example, the use of individual LEDs emitting RGB light and Ce:YAG LEDs, which convert blue light into yellow light, and the mix of these colors generates white light [51]. This spectral engineering approach demonstrates that a perfect balance between the relative intensities of individual light emissions is desirable. Concerning REI-doped glasses for white light generation, there are several combinations of REI and excitation methods: Dy^{3+}, Gd^{3+}-Dy^{3+}, and Tb^{3+}-Dy^{3+} upon UV excitation [52–54]; Er^{3+}-Pr^{3+} upon 450 nm excitation [55]; Er^{3+}-Dy^{3+}-Sm^{3+} and Er^{3+}-Tm^{3+}-Sm^{3+} upon UV excitation [56, 57], or Er^{3+}-Yb^{3+}-Tm^{3+} via UC [58]. The mentioned systems enable tunable white light across the CCT, i.e., warm white light (below 3700 K), artificial daylight (4000 K), sunlight (5500 K), and cold white light (larger than 6000 K) [59, 60]. This tailoring is achieved by varying the excitation power or the concentration of the REI. This approach offers a distinct advantage over conventional white LEDs, as it allows creation of photonic devices capable of on-demand tuning to preferred white light tones.

Zinc-tellurite glasses doped with 3 mol% of Yb^{3+}, 0.1 mol% of Tm^{3+}, and 0.05 mol% of Er^{3+} were fabricated, and we labeled as ErYbTm. The selected concentration of REI is based on the sample composition given in Ref. [59], where Lozano C. et al. obtained cold white light to blue emission varying the excitation power. Here, we demonstrated that with a slight increase in Er^{3+} content we can expand the white light tones. First, the UC spectra of the ErYbTm sample under different excitation powers is depicted in Fig. 3.8a. We can observe emission bands centered at 472, 651, and 692 nm, ascribed to the Tm^{3+}: $^1G_4 \rightarrow {}^3H_6$, $^5F_4 \rightarrow {}^3H_6$, and $^3F_3 \rightarrow {}^3H_6$ transitions, respectively. Moreover, the bands peaked at 521, 544 and a shoulder at 662 nm correspond to the Er^{3+}: $^2H_{11/2} \rightarrow {}^4I_{15/2}$, $^4S_{3/2} \rightarrow {}^4I_{15/2}$, and $^4F_{9/2} \rightarrow {}^4I_{15/2}$ transitions, respectively. Subtle steps of pump power increments (5 mW) are chosen to ensure a uniform evolution in the white light tones. Besides, at higher excitation powers, the relative intensities of the blue and red emission bands are approximately the same. This could generate purple light emission in the CIE 1931 chromaticity diagram; however, the green emission of Er^{3+} contributes to tilting the chromaticity trend toward green. This effect is indeed observed in the chromaticity diagram displayed in Fig. 3.8b, where the closest point to the (0.333, 0.333) coordinate is (0.325, 0.332), which corresponds to the spectra at 30 mW of excitation power (see the inset), while at 20 and 25 mW are the nearest to a CCT of 4500 K. Further, the increment in excitation power leads to an intense blue emission, and it may be due to enhanced ET from Er^{3+} to Tm^{3+} or to a luminescence quenching of Er^{3+} at higher pump powers. These statements can be verified through lifetime measurements. Finally, we conclude that a tunable white light source can be fabricated using REI-doped tellurite glasses by varying the excitation power.

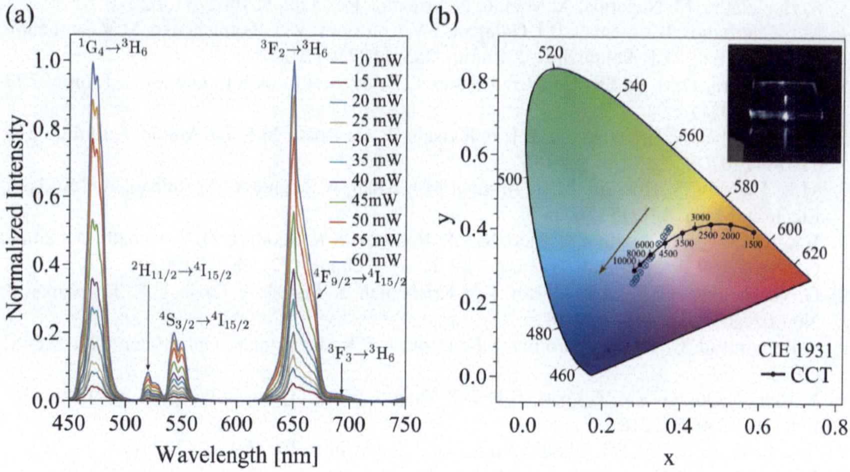

Fig. 3.8 **a** UC emission spectra of the ErYbTm sample upon 980 nm excitation at different pump powers. **b** CIE 1931 chromaticity diagram of the UC spectra. The arrow indicates the increment in excitation power. *Inset*: Photograph of the ErYbTm emitting white light under 30 mW, captured using a conventional smartphone camera

Summarizing, we first explored the fundamentals of detection and emission of light in REI, providing discussion about different emission mechanisms, i.e., down-shifting, down-conversion, quantum cutting, and up-conversion. Both excitation and emission spectroscopy were analyzed with a Pr^{3+}-doped tellurite glass, where we constructed its energy level diagram. Further, the up-conversion process was briefly discussed and a complete spectroscopic analysis of an Er^{3+}-doped tellurite glass upon 980 nm excitation was given. This analysis considers the distinction between using the peak maximum position and the integrated band areas, when calculating the number of 980 nm photons involved in achieving population inversion within the up-conversion emission bands. Regarding the emission cross-section, we discussed about the two main methods employed in its calculation: McCumber and Füchtbauer–Ladenburg relations. The discrepancies between these two methods are highlighted with experimental evidence. Finally, colored and white light emissions in glasses doped with different REI (doped and co-doped) were explored according to the CIE 1931 standards. In this manner, we have outlined a comprehensive framework for the emission analysis of REI-doped glasses, offering both theoretical insights and experimental guidance.

References

1. P. Singh, P. Singh, S.K. Singh, in *Modern Techniques of Spectroscopy*, ed. by D.K. Singh, M. Pradhan, A. Materny (Springer Singapore, Singapore, 2021), pp. 389–407
2. I.C. Pinto, V.A.G. Rivera, R. Falci, V. Fuertes, S. LaRochelle, Y. Messaddeq, J. Lumin. **265**, 120206 (2024)
3. S. Hemalatha, M. Nagaraja, A. Madhu, N. Srinatha, Res. Opt. **9**, 100275 (2022)
4. M.F. Churbanov, B.I. Denker, B.I. Galagan, V.V. Koltashev, V.G. Plotnichenko, M.V. Sukhanov, S.E. Sverchkov, A.P. Velmuzhov, J. Lumin. **245**, 118756 (2022)
5. J.G. Câmara, D.M. da Silva, L.R.P. Kassab, C.B. de Araújo, A.S.L. Gomes, J. Lumin. **233**, 117936 (2021)
6. D.S. Da Silva, N. Ursus Wetter, L.R.P. Kassab, W. De Rossi, M.S. De Araujo, J. Lumin. **217**, 116789 (2020)
7. M.K. Hossain, S. Hossain, M.H. Ahmed, M.I. Khan, N. Haque, G.A. Raihan, A.C.S. Appl, Electron. Mater. **3**, 3715 (2021)
8. V.K. Kummara, G. Neelima, N. Ravi, N.N.K. Reddy, H.S.K. Reddy, C.D. Viswanath, D. Lenine, G. Surekha, R.P. Suvarna, C. Yuvaraj, Opt. Laser Technol. **127**, 106167 (2020)
9. G. Venkataiah, P. Babu, I.R. Martín, K.V. Krishnaiah, K. Suresh, V. Lavín, C.K. Jayasankar, J. Non-Cryst. Solids **479**, 9 (2018)
10. P. Ramprasad, C. Basavapoornima, S.R. Depuru, C.K. Jayasankar, Opt. Mater. **129**, 112482 (2022)
11. Y. Ren, Z. Qin, G. Xie, Z. Qiao, T. Hai, P. Yuan, J. Ma, L. Qian, S. Wang, C. Yu, Chin. Opt. Lett. **16**, 020020 (2018)
12. R.S. Yadav, S.B. Rai, S.J. Dhoble, Prog. Solid State Chem. **57**, 100267 (2020)
13. M. Seshadri, I.T.C. Santos, M.J.V. Bell, J. Lapointe, Y. Messaddeq, V. Anjos, Sci. Rep. **12**, 19311 (2022)
14. H. Singh, T. Singh, D. Singh, V. Bhatia, D. Kumar, S.P. Singh, Opt. Mater. **137**, 113586 (2023)
15. F. Enrichi, E. Cattaruzza, P. Riello, G.C. Righini, A. Vomiero, Ceram. Int. **47**, 17939 (2021)
16. A. Chiappini, L. Zur, F. Enrichi, B. Boulard, A. Lukowiak, G.C. Righini, M. Ferrari, *Solar Cells and Light Management* (Elsevier, 2020), pp. 391–414

17. A.R. Khadiev, S.L. Korableva, A.K. Ginkel, O.A. Morozov, A.S. Nizamutdinov, V.V. Semashko, M.S. Pudovkin, Opt. Mater. **134**, 113118 (2022)
18. N.S. Satpute, C.M. Mehare, A. Tiwari, H.C. Swart, S.J. Dhoble, A.C.S. Appl, Electron. Mater. **4**, 3354 (2022)
19. M. Seshadri, M.J.V. Bell, V. Anjos, Y. Messaddeq, J. Rare Earths **39**, 33 (2021)
20. M.K. Poojha, K.A. Naseer, P. Matheswaran, K. Marimuthu, E. El Shiekh, Opt. Mater. **143**, 114289 (2023)
21. G.C. Lozano, J. Chacaliaza-Ricaldi, J.L.H. Clabel, Y. Messaddeq, V.A.G. Rivera, E. Marega, J. Lumin. **260**, 119853 (2023)
22. C. Basavapoornima, C.R. Kesavulu, T. Maheswari, W. Pecharapa, S.R. Depuru, C.K. Jayasankar, J. Lumin. **228**, 117585 (2020)
23. F. Auzel, C.R. Acad, Sci. Paris **262**, 1016 (1966)
24. R. Miedzinski, I. Fuks-Janczarek, L.R.P. Kassab, F.A. Bomfim, Mater. Res. Bull. **95**, 339 (2017)
25. M. Dhavamurthy, P. Vinothkumar, M. Mohapatra, A. Suresh, P. Murugasen, Spectrochim. Acta Part A Mol. Biomol. Spectrosc. **266**, 120448 (2022)
26. F. Auzel, J. Lumin. **223**, 116900 (2020)
27. G.C. Lozano, V.A.G. Rivera, Y. Messaddeq, E.M. Jr., J. Alloys Compd. **948**, 169777 (2023)
28. A. Kumar, A. Bahadur, J. Alloy. Compd. **857**, 158196 (2021)
29. J.F. Suyver, A. Aebischer, S. García-Revilla, P. Gerner, H.U. Güdel, Phys. Rev. B **71**, 125123 (2005)
30. H. Kumari, G.F. Ansari, S.K. Mahajan, K. Sk Rezaul, S. Bairagi, Mater. Today Proc. (2023)
31. D. Manzani, J.F.S. da Petruci, K. Nigoghossian, A.A. Cardoso, S.J.L. Ribeiro, Sci. Rep. **7**, 41596 (2017)
32. L.N. Lindolfo da Silva, A. Torquato, I. Nunes de Assis, M.R. Dousti, Solid State Sci. **112**, 106444 (2021)
33. O.B. Silva, V.A.G. Rivera, Y. Ledemi, Y. Messaddeq, E. Marega, J. Lumin. **232**, 117808 (2021)
34. P. Vani, G. Vinitha, N. Manikandan, J. Mater. Sci. Mater. Electron. **34**, 32 (2023)
35. G. Eryurek, S. Tabanli, T. Buhari, M. Erdem, ECS J. Solid State Sci. Technol. **12**, 076005 (2023)
36. M. Chaika, S. Balabanov, D. Permin, Opt. Mater. **112**, 110785 (2021)
37. V.A.G. Rivera, I.C. Pinto, R. Falci, V. Fuertes, Y. Messaddeq, J. Alloy. Compd. **976**, 173182 (2024)
38. H.C. Çamiçi, T. Guérineau, V.A.G. Rivera, R.F. Falci, S. LaRochelle, Y. Messaddeq, Sci. Rep. **13**, 8835 (2023)
39. D.E. McCumber, Phys. Rev. **136**, A954 (1964)
40. Ł Marek, M. Sobczyk, J. Non-Cryst. Solids **576**, 121238 (2022)
41. P. Meejitpaisan, H.J. Kim, J. Kaewkhao, Integr. Ferroelectr. **239**, 104 (2023)
42. D. Rich, in *Fundamentals and Applications of Color Engineering*, 1st edn., ed. by P. Green (Wiley, 2023), pp. 1–25
43. M. Magnusson, J. Sigurdsson, S.E. Armansson, M.O. Ulfarsson, H. Deborah, J.R. Sveinsson, in *IGARSS 2020—2020 IEEE International Geoscience and Remote Sensing Symposium* (2020), pp. 2045–2048
44. J. Zwinkels, in *Encyclopedia of Color Science and Technology*, ed. by R. Shamey (Springer International Publishing, Cham, 2023), pp. 110–114
45. G. Pullaiah, K.V. Rao, B.C. Jamalaiah, A.S.N. Reddy, J. Non-Cryst. Solids **615**, 122413 (2023)
46. V. Hegde, K.R. Vighnesh, S.D. Kamath, C.S.D. Viswanath, A.H. Almuqrin, M.I. Sayyed, J. Gangareddy, R. Rajaramakrishna, K. Keshavamurthy, Appl. Phys. A **130**, 411 (2024)
47. E.M. Bouabdalli, M. El Jouad, N. Gaumer, S. Touhtouh, A. Hajjaji, J. Lumin. **256**, 119640 (2023)
48. O.C. Silva Neto, N.F. Dantas, T.A. Lodi, J.G. Oliveira Neto, T.S. Morais, F. Pedrochi, A. Steimacher, Opt. Mater. **154**, 115761 (2024)
49. N. Wantana, E. Kaewnuam, Y. Ruangtaweep, D. Valiev, S. Stepanov, K. Yamanoi, H.J. Kim, S. Kothan, J. Kaewkhao, J. Non-Cryst. Solids **554**, 120603 (2021)
50. A. Ruivo, C. Laia, Color. Res. Appl. **49**, 199 (2024)

51. H. Yang, Y. Zhang, Y. Zhang, Y. Zhao, X. Liang, G. Chen, Y. Liu, W. Xiang, Chem. Eng. J. **414**, 128754 (2021)
52. M. Monisha, N. Mazumder, G. Lakshminarayana, S. Mandal, S.D. Kamath, Ceram. Int. **47**, 598 (2021)
53. W. Rittisut, N. Wantana, Y. Ruangtaweep, P. Mool-am-kha, J. Padchasri, S. Rujirawat, P. Manyum, R. Yimnirun, P. Kidkhunthod, A. Prasatkhetragarn, S. Kothan, H.J. Kim, J. Kaewkhao, Opt. Mater. **122**, 111705 (2021)
54. L. Vijayalakshmi, K. Naveen Kumar, J.D. Baek, J. Rare Earths **42**, 46 (2024)
55. J. Pisarska, M. Kuwik, A. Baranowska, M. Kochanowicz, J. Żmojda, P. Miluski, J. Dorosz, M. Leśniak, B. Starzyk, D. Dorosz, W.A. Pisarski, Mater. Res. Bull. **150**, 111791 (2022)
56. A. Jose, K. T, A. George, T.A. Jose, C. Joseph, N.V. Unnikrishnan, P.R. Biju, J. Non-Cryst. Solids **562**, 120775 (2021)
57. T.O. Sales, C. Jacinto, W.F. Silva, R. Antunes, D.T. Dias, A. Gonçalves, R. El-Mallawany, N.G.C. Astrath, A. Novatski, J. Alloy. Compd. **899**, 163305 (2022)
58. H. Liang, W. Lei, S. Liu, P. Zhang, Z. Luo, A. Lu, Opt. Mater. **119**, 111320 (2021)
59. G. Lozano, O.B. Silva, W. Faria, A.S.S. de Camargo, R.R. Gonçalves, D. Manzani, R. Bruna, V.A.G. Rivera, E. Marega, J. Lumin. **228**, 117538 (2020)
60. G. Mahajan, M.V.V.K.S. Prasad, K. Swapna, K. Seshulatha, G.G. Dhar, A.S. Rao, Opt. Mater. **149**, 115067 (2024)

Chapter 4
Decay Lifetime Curves

Analysis of decay lifetime curves is a fundamental tool for understanding the photophysical behavior of REI-doped materials [1–4]. These measurements provide direct insight into the temporal dynamics of excited states, allowing the determination of radiative and non-radiative decay rates, as well as the efficiency of energy transfer processes [5, 6]. In REI, which typically exhibit long-lived excited states due to parity-forbidden *f–f* transitions, lifetime measurements are particularly valuable for characterizing the local environment of the ion and assessing quenching mechanisms induced by defects, phonons, or high dopant concentrations. Furthermore, decay curves can reveal multi-exponential behavior indicative of energy transfer or migration effects, making them indispensable for the development and optimization of optical materials for applications such as lasers, phosphors, and up-conversion devices.

In Chap. 2, various mechanisms by which the luminescent system can absorb excitation energy are discussed. Chapter 3 addressed several possible pathways for returning to the ground state. In the present chapter, we focus on the temporal behavior of metastable levels of REI; that is, we present and discuss in detail the role of radiative processes in the fluorescence of ion-doped matrices. NR transitions, which are the dominant decay mechanism between energetically close electronic levels in the ions, will be addressed in Chap. 6.

4.1 Measurement of Decay Time

Decay time measurements enable the investigation of both radiative and non-radiative processes associated with the energy levels of REIs. A typical experimental setup consists of an optical excitation system and a luminescence detection unit coupled with a time-resolved acquisition system capable of monitoring the transient response following the interruption of excitation.

G. Lozano C. et al., *A Brief Guide to Optical Spectroscopy of Trivalent Rare Earth Ions in Glasses*, Springer Series in Optical Sciences 247,
https://doi.org/10.1007/978-3-032-18501-3_4

Excitation is commonly achieved using a laser source tuned to a specific absorption band of the ions of interest. To resolve the temporal profile of the emission, the excitation must be pulsed. This can be accomplished with commercially available pulsed lasers or, alternatively, by modulating continuous-wave lasers using mechanical devices such as optical choppers. This approach is particularly feasible because of the relatively long radiative lifetimes of REI, which typically range from microseconds to milliseconds, allowing for the use of low-frequency modulation.

The emitted luminescence is collected by an optical assembly and directed to a sensitive detector that records the emission intensity as a function of time. Photomultiplier tubes (PMTs) are traditionally employed for this purpose, because of their high sensitivity and fast temporal response, which makes them ideal for detecting low-intensity signals with high time resolution. Alternatively, semiconductor photodetectors are also widely used. Silicon (Si) photodetectors offer good spectral sensitivity in the visible range up to approximately 1000 nm, while indium gallium arsenide (InGaAs) detectors are suitable for detecting near-infrared luminescence, extending up to about 1700 nm, depending on the model.

The detected signal is typically amplified and analyzed using a high-resolution digital oscilloscope or a data acquisition system synchronized with the excitation pulse. From the acquired data, decay curves can be fitted with exponential functions. In cases where the relaxation process is not influenced by additional mechanisms—such as energy transfer or reabsorption, the decay profile follows a single-exponential behavior, with a time constantly corresponding to the intrinsic lifetime of the excited state.

4.2 Single-Exponential Decay

Considering the two-level diagram in Fig. 4.1, the probability of spontaneous emission (W^{rad}) derived by Einstein for a radiative transition between the initial state $|a\rangle$ and final state $|b\rangle$, assuming an electric dipole interaction, is given by

$$W^{rad}(\nu) = \frac{1}{4\pi\varepsilon_0}\frac{8\pi^2 e^2 \nu^2}{mc^3}\left[\left(\frac{n^2+2}{3}\right)^2 n\right] f(ab) \tag{4.1}$$

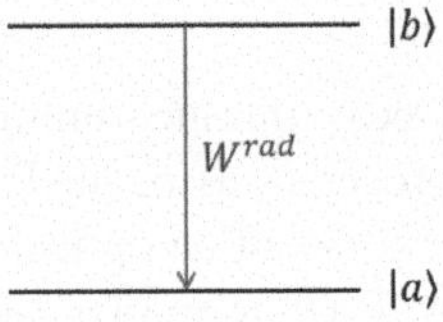

Fig. 4.1 Schematic representation of radiative decay in a two-level system

where n is the refractive index, m is the electron mass, e is the elementary charge, v is the transition frequency, c is the speed of light, and $f(ab)$ is the oscillator strength, defined as

$$f(ab) = \frac{1}{g_a}\sum_{a,b}\frac{8\pi^2 mv}{3he^2}|\langle b|\boldsymbol{D}|a\rangle|^2 \tag{4.2}$$

where g_a is the degeneracy of the initial state, h is Planck's constant, and $\boldsymbol{D}$ is the electric dipole moment operator.

The probability of spontaneous emission is related to the radiative lifetime (τ_{rad}) of the excited state as follows:

$$W^{rad} = \frac{1}{\tau_{rad}} = \sum_{ab} A_{ab} \tag{4.3}$$

where A_{ab} is given by

$$A_{ab} = \frac{8\pi cn^2}{\lambda_{ij}^4 N}\frac{g_b}{g_a}\int \alpha(\lambda)d\lambda = \frac{8\pi cn^2\sigma_{em}(\lambda)}{\lambda_{ab}^4 P(\lambda)} \tag{4.4}$$

where g_b and g_a are the degenerations of the upper and lower levels, respectively, $\alpha(\lambda)$ is the absorption coefficient (discussed in Chap. 2), N is the dopant ion concentration (in cm^{-3}), λ_{ab} is the wavelength corresponding to the maximum absorption for the transition between $|a\rangle$ and $|b\rangle$ states, $\sigma_{em}(\lambda)$ is the emission cross-section, and $P(\lambda)$ is the function of the normalized line profile of the fluorescence.

It should be noted that τ_{rad} values for allowed transitions are typically in the order of nanoseconds. For instance, τ_{rad} of sodium in the $^3P_{3/2} \rightarrow {}^3S_{1/2}$ ($\lambda = 589.1$ nm) transition is $\tau_{rad} = 16$ ns. On the other hand, forbidden transitions have long lifetimes, as can be seen in the 2S $\rightarrow$ 1S transition of hydrogen atom that has $\tau_{rad} = 1$ s (emission allowed by two photons). In the case of TR^{3+}, we discussed that, in general, transitions are forbidden. However, the crystalline field promotes a mix of parties making them allowed by forced electric dipole mechanisms. So, the lifetimes of these transitions are in the order of milliseconds.

In the case of REIs, when the excited-state transitions do not undergo significant influence from energy transfer processes—such as cross-relaxation, up-conversion, or energy migration—the decay of the luminescence can be well described by a single-exponential function. This behavior is characteristic of an isolated radiative or non-radiative relaxation channel, where the excited-state population decays independently and with a well-defined lifetime.

The luminescence intensity $I(t)$ as a function of time can be expressed as follows:

$$I(t) = I(0)e^{-t/\tau} \tag{4.5}$$

where $I(0)$ is the initial intensity at $t = 0$ and τ is the intrinsic lifetime of the excited state. This lifetime reflects the probability of spontaneous emission or non-radiative relaxation events, depending on the specific energy level and host matrix.

Single-exponential decay profiles are often observed in low-concentration samples, where rare-earth ions are sufficiently far apart to prevent interaction, or in matrices specifically designed to suppress parasitic energy transfer. Under these conditions, τ provides direct information about the fundamental de-excitation mechanisms of the ion in the given environment, enabling comparison across different host materials and aiding in the evaluation of quantum efficiency.

It is important to note that, in the absence of energy transfer processes or other non-radiative mechanisms, the intrinsic lifetime τ introduced in Eq. (4.5) corresponds to the radiative lifetime τ_{rad} defined in Eq. (4.3). In this ideal case, where the excited-state population decays solely via spontaneous emission, the total decay rate is governed exclusively by the radiative transition probabilities A_{ab}, and non-radiative losses can be neglected. As a result, the experimentally observed single-exponential decay directly reflects the fundamental radiative properties of the transition and the local photonic environment, enabling a straightforward comparison with theoretical predictions derived from Einstein's coefficients. For instance, Fig. 4.2 illustrates the step-by-step process for determining the single-exponential lifetime (τ) of the excited state $^5D_4$ of the Tb^{3+}-doped low silica calcium aluminosilicate glass (LSCAS), under excitation at 488 nm [7].

As shown in Fig. 4.2a, excitation at a wavelength of 488 nm ($\lambda_{ex} = 488$ nm) promotes electrons from the ground state $^7F_6$ to the excited $^5D_4$ level. From this level, radiative transitions can occur to various lower-energy levels of the ^7F_J multiplet (J = 0–6), as indicated in Fig. 4.2a. Among these transitions, the emission from $^5D_4$ to $^7F_5$ (545 nm) is the most intense and is monitored as a function of time.

In Fig. 4.2b, the decay curve $I(t)$ of $^5D_4 \rightarrow {}^7F_5$ metastable level is recorded experimentally. The luminescence intensity decreases over time as a result of the de-excitation of the excited-state population. This temporal evolution can be modeled by the single-exponential function given by Eq. (4.5). To better visualize the exponential decay and facilitate normalization, the intensity data are commonly plotted as $I(t)/I(0)$ versus time, as illustrated in Fig. 4.2c. This normalized representation helps emphasize the functional form of the decay, independent of the absolute signal amplitude.

For a more accurate determination of the lifetime τ, the logarithmic form of Eq. (4.5) is often used $\mathrm{Log}[I(t)/I(0)] = -t/\tau$. By plotting the natural logarithm of the normalized intensity versus time, a linear relationship is expected for single-exponential decay processes. As depicted in Fig. 4.2d, the data indeed follow a straight line, and the slope of this line ($-1/\tau$) allows direct extraction of the lifetime, $\tau = (1.90 \pm 0.02)$ ms, through linear regression.

This methodology provides a straightforward and robust way to quantify the radiative properties of REI-doped material. In the absence of non-radiative processes or interactions between ions, the extracted lifetime τ corresponds to the radiative lifetime τ_{rad} introduced in Eq. (4.3).

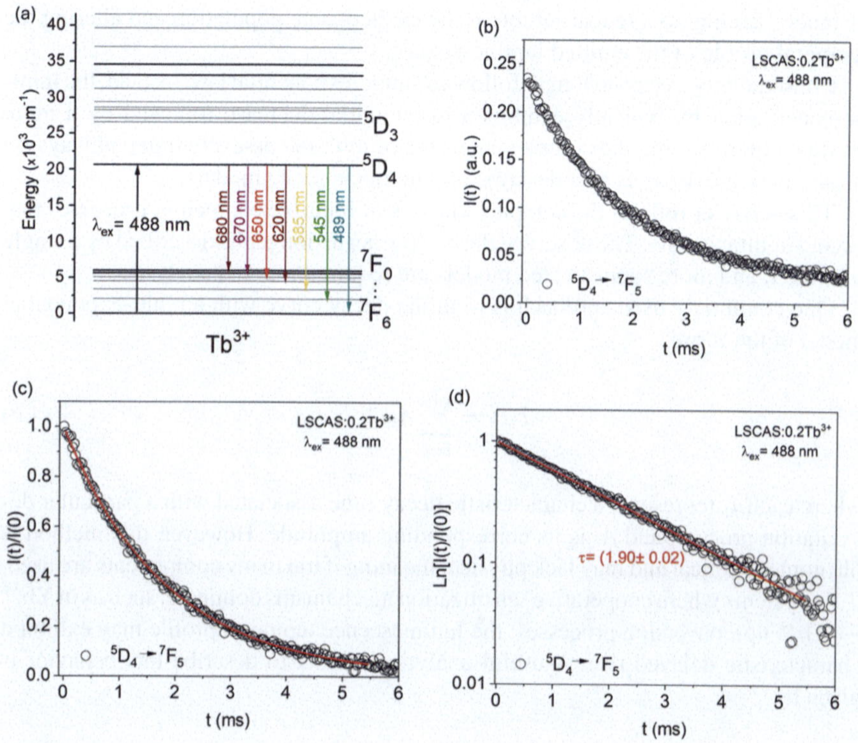

Fig. 4.2 **a** Energy level diagram of Tb^{3+} ions in the LSCAS host, indicating excitation at 488 nm and subsequent $^5D_4 \rightarrow {}^7F_{0\text{-}6}$ emission. **b** Experimental luminescence decay curve $I(t)$ of the $^5D_4 \rightarrow {}^7F_{0\text{-}6}$ transition under 488 nm. **c** Normalized decay curve $I(t)/I(0)$, emphasizing the exponential behavior. **d** Natural logarithm of the normalized intensity as a function of time used to extract the single-exponential lifetime $\tau = (1.90 \pm 0.02)$ ms from the slope of the linear fit

4.3 Non-exponential Decay

The single-exponential behavior described in Sect. 4.2 is generally observed for excited states that are not significantly involved in energy transfer processes. In particular, this applies to energy levels that are sufficiently isolated—both spectrally and spatially—from other ions or quenching centers, so that the decay of the excited-state population occurs predominantly via intrinsic radiative channels. This condition is often met in low-concentration samples or when the excited state corresponds to a level that is energetically and spectroscopically "protected," i.e., it is not resonant with levels of neighboring ions that would allow transfer processes to occur.

However, in many practical systems, this ideal scenario is not sustained. As the dopant concentration increases or depends on the characteristics host matrix, the likelihood of energy transfer mechanisms such as cross-relaxation, energy migration, or up-conversion becomes significant. These processes open additional decay channels, leading to a redistribution of the excited-state population and altering the temporal profile of the emitted luminescence.

Consequently, decay no longer follows a single-exponential law. Instead, the luminescence intensity typically exhibits a faster initial decrease followed by a more gradual tail, reflecting the complex interplay of multiple de-excitation pathways. In these cases, the decay is best described by more elaborate models.

This behavior reflects the complex kinetics of multiple competing processes that occur simultaneously. The observed decay $I(t)$ can no longer be described by a single lifetime τ, and more sophisticated models are required for data analysis.

One commonly used approach is to fit the decay curve with a multi-exponential model of the form:

$$I(t) = \sum_{i=1}^{n} A_i e^{-t/\tau_i} \tag{4.6}$$

where each τ_i represents a characteristic decay time associated with a particular de-excitation process, and A_i is its corresponding amplitude. However, this method is phenomenological and may lack physical meaning if too many components are used.

In systems where cooperative sensitization mechanisms dominate, such as in $Yb^{3+} \rightarrow Tb^{3+}$ up-conversion processes, the luminescence temporal profile may exhibit a characteristic delayed rise. A useful analytical model to describe this behavior is given by

$$I(t) = A\left(e^{-Bt} - e^{-Ct}\right) \tag{4.7}$$

where A is a normalization constant, B and C are kinetic parameters related to the rise and decay rates, respectively.

This expression captures both the rise of the luminescence signal due to population build-up of the excited state via energy transfer and its subsequent radiative decay. The intensity initially increases as the population accumulates (dominated by the slower exponential term), reaching a maximum when the rates balance, and then decreases exponentially. This profile is typical when the population of the emitting level is governed by an intermediate reservoir, as expected for cooperative energy transfer between two excited Yb^{3+} ions to a single Tb^{3+} ion.

This model has been successfully applied, for instance, in the time-resolved study of $SrCl_2$:Tb^{3+}/Yb^{3+} [8] systems under pulsed excitation, providing evidence that the up-conversion mechanism proceeds via cooperative sensitization rather than excited-state absorption.

Alternatively, stretched exponential functions (also known as Kohlrausch–Williams–Watts functions) [9] are often employed in systems where energy transfer is governed by a distribution of distances or interaction strengths [9, 10]:

$$I(t) = I(0)e^{(-t/\tau_i)^{\beta}} \tag{4.8}$$

where $0 < \beta < 1$ is the stretch parameter, which accounts for the broadening of the decay kinetics due to disorder or inhomogeneity in the system.

In the specific case of concentration quenching, where high dopant concentrations lead to increased ion–ion interactions, the lifetime becomes strongly dependent on dopant density. Comparison of decay curves as a function of concentration can provide insight into the efficiency and range of energy transfer mechanisms.

Additionally, Förster or Dexter models can be used to extract transfer rates and critical distances, depending on whether dipole–dipole or exchange interactions dominate.

Therefore, when non-exponential decay is observed, it is essential to go beyond the single lifetime approximation and adopt appropriate models that reflect the underlying physical processes. The choice of model should be guided by the nature of the energy transfer mechanism; the dimensionality of the system; and experimental parameters such as temperature, excitation power, and dopant concentration.

The following examples illustrate how non-exponential decay manifests in different rare-earth-doped systems:

- **Nd^{3+}-Doped Materials**: Neodymium ions exhibit strong energy migration among identical ions through resonant dipole–dipole interactions [11, 12]. In concentrated samples, cross-relaxation processes from the $^4F_{3/2}$ excited state to the ground state $^4I_{15/2}$ led to quenching of luminescence and complex decay profiles. These effects intensify with increasing concentration of Nd^{3+}, and lifetime measurements as a function of concentration can reveal critical transfer distances.
- **Tb^{3+}-Doped Phosphors**: In materials doped with Tb^{3+}, the $^5D_3$ level is particularly sensitive to cross-relaxation processes [1, 13]. These interactions typically become significant at concentrations above 0.5 mol%, depending on the host matrix. The cross-relaxation mechanism, such as $^5D_3 + {}^7F_6 \rightarrow {}^5D_4 + {}^7F_0$, transfers excitation energy between neighboring Tb^{3+} ions, depopulating the $^5D_3$ level in favor of $^5D_4$, since the energy steps $^5D_3 \rightarrow {}^5D_4$ (5800 cm^{-1}) and $^7F_6 \rightarrow {}^7F_0$ (5700 cm^{-1}) are remarkably similar. As a result, the blue emission from $^5D_3 \rightarrow {}^7F_{J=3\text{–}6}$ transitions is suppressed, the green emission from $^5D_4 \rightarrow {}^7F_5$ becomes dominant, and the luminescence decay evolves from single- to multi-exponential. The decay time of the $^5D_3$ level decreases with increasing concentration, reflecting the enhanced probability of non-radiative energy transfer.
- Blue curve in Fig. 4.3a shows the $^5D_3 \rightarrow {}^7F_6$ decay curve corresponding to higher Tb^{3+} concentration (2.0 wt%) in Tb^{3+}-doped LSCAS glass. It decreases more rapidly compared to the curve of the sample with lower concentration (0.2 wt%) in the same glass. This behavior reflects the increased efficiency of cross-relaxation processes, which leads to a faster depopulation of the $^5D_3$ level and a consequent population build-up of $^5D_4$ level.
- This difference becomes even more evident in Fig. 4.3b, where the experimental data are plotted on a semi-logarithmic scale. Although the sample with lower Tb^{3+} concentration exhibits a typical single-exponential decay behavior—indicating negligible energy transfer—the higher concentration sample clearly deviates from linearity, revealing the multi-exponential nature of the decay. This pronounced

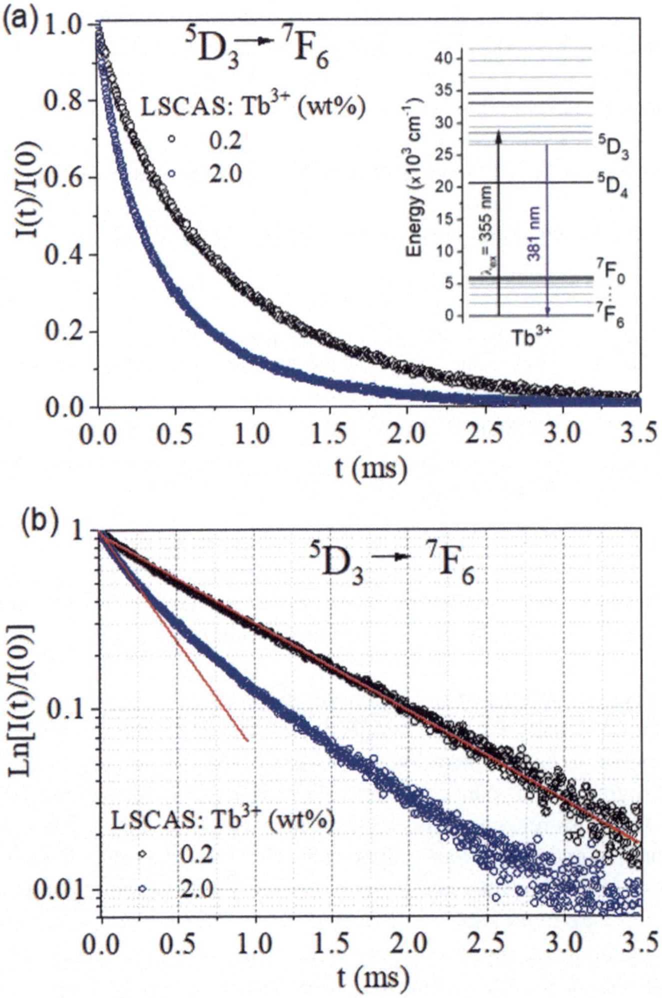

Fig. 4.3 **a** Luminescence decay curves of the $^5D_3 \rightarrow {}^7F_6$ transition for Tb^{3+}-doped LSCAS glasses with concentrations of 0.2 wt% (black) and 2.0 wt% (blue), under 355 nm excitation. A faster decay is observed for the higher Tb^{3+} concentration, indicating enhanced non-radiative energy transfer due to cross-relaxation. **b** The same decay curves presented on a semi-logarithmic scale. The curve for 0.2 wt% Tb^{3+} exhibits a typical single-exponential behavior, whereas the deviation from linearity in the 2.0 wt% sample reveals a pronounced multi-exponential profile, confirming the presence of cross-relaxation processes

curvature is a clear signature of the occurrence of ET processes, which introduce additional NR de-excitation channels. Therefore, the comparison between these two samples demonstrates how increasing the Tb^{3+} concentration enhances ion–ion interactions, significantly impacting both the emission profile and the luminescence decay kinetics.

- Yb^{3+}-Doped Systems: Although Yb^{3+} has only two levels, energy migration among Yb^{3+} ions can occur at high concentrations [14]. This migration may lead to quenching at defect sites or surface states, resulting in reduced lifetimes and stretched exponential decay behavior, especially in nanocrystals or disordered hosts.
- Er^{3+}–Yb^{3+} Co-doped Systems: Yb^{3+} acts as an efficient sensitizer by absorbing near-infrared light (e.g., 980 nm) and transferring energy to Er^{3+} [2, 15]. This donor–acceptor interaction leads to population of higher Er^{3+} levels (e.g., $^4I_{11/2}$), which subsequently decay to the metastable $^4I_{13/2}$ state. The resulting Er^{3+} emission at 1.53 μm often displays a characteristic rise time due to the energy transfer dynamics, followed by a decay that may include multiple components. Accurate modeling requires coupled rate equations that describe the Er^{3+}–Yb^{3+} interactions.
- Tm^{3+}–Yb^{3+} Systems: In up-conversion materials, Yb^{3+} sensitization enables population of Tm^{3+} excited states through sequential energy transfers. These processes give rise to multi-step mechanisms that result in non-exponential emission decays in the blue and UV spectral regions. The complexity of the decay increases with excitation power and Tm^{3+} concentration, reflecting the nonlinear nature of the up-conversion pathways [6].

These examples highlight the importance of considering energy transfer mechanisms when analyzing luminescence decay data. Non-exponential decay is not a limitation, but rather a rich source of information on ion–ion interactions, material homogeneity, and the efficiency of radiative processes in rare-earth-doped systems.

4.4 Inokuti–Hirayama Model

The purpose of this section is to present models of non-exponential fluorescence. These models, which measure the time evolution of the fluorescence amplitude of various luminescent materials, are important in fields such as solid-state physics and organic chemistry, as well as in optical engineering [16–18].

Much of the interest in non-exponential fluorescence is due to the early work of Förster [19], Dexter [20], Inokuti and Hirayama [21]. The theories of these researchers have been applied for more than 70 years.

The Inokuti–Hirayama (IH) model, proposed in 1965, is a foundational theoretical framework used to describe non-exponential decay behavior in luminescent materials where energy transfer occurs from excited donor ions to nearby acceptor ions. This model is particularly relevant for systems containing RE dopants, where concentration quenching and interaction mechanisms such as dipole–dipole, dipole–quadrupole, or exchange interactions dominate the luminescence dynamics. When a luminescent ion (donor) is excited, it can either return to the ground state via radiative decay, or transfer energy to a neighboring acceptor ion, which may then relax non-radiatively [22].

Table 4.1 Interaction types

Interaction type	s value
Dipole–dipole	6
Dipole–quadrupole	8
Quadrupole–quadrupole	10

This donor–acceptor interaction modifies the expected exponential luminescence decay law. In systems with random spatial distribution of acceptors, the overall decay becomes non-exponential [23, 24]. The luminescence intensity $I(t)$ as a function of time is given by [21]

$$I(t) = I_0 \cdot \exp\left(-\frac{t}{\tau_0}\right) \cdot \exp\left(-Q\left(\frac{t}{\tau}\right)^{3/s}\right) \tag{4.9}$$

where I_0 is the initial amplitude for the fluorescence, τ_0 is the characteristic lifetime of the exponential fluorescence decay that would result if there were no energy transfer between the excited states of atomic electrons in the RE-doped glass, the dimensionless scaling factor Q is the energy transfer parameter (depends on acceptor concentration and interaction strength), and the exponent s determines the nature of the involved electric multipole interactions. This model results from the energy transfer [19] between two populations of excited states during the fluorescence time [25]. The value of s is determined by the nature of the donor–acceptor coupling, see Table 4.1.

The parameter Q is defined as follows:

$$Q = \frac{4}{3\pi^{3/2}}\Gamma\left(1 - \frac{3}{s}\right)N_0R_0^3 \tag{4.10}$$

where Γ is the Euler function, N_0 is the concentration of acceptor ions, and R_0 is the critical transfer distance defined as a donor–acceptor separation for which the rate of energy transfer between a donor–acceptor is equal to the rate of intrinsic decay rate τ_0^{-1} [26, 27]. Using the model defined in Eq. (4.8) to fit the decay curves, the nature of the interaction involved in the ET may be accessed, starting by normalizing the fluorescence intensity $I(t)/I_0$ and applying the logarithm:

$$\ln\left(\frac{I(t)}{I_0}\right) = -\frac{t}{\tau_0} - Q\left(\frac{t}{\tau}\right)^{3/s} \tag{4.11}$$

$$-\ln\left(-\ln\left(\frac{I(t)}{I_0}\right) - \frac{t}{\tau_0}\right) = \ln Q + \frac{1}{s}\ln\left(\frac{t}{\tau_0}\right)^3 \tag{4.12}$$

where Eq. (4.12) should yield a straight line with slope $3/s$ [1].

The IH model is especially useful in rare-earth-doped glasses, where the spectral overlap between donor emission and acceptor absorption exists and the dopant concentration is high enough to promote energy migration and quenching effects become significant at higher concentrations [28]. By fitting the experimental decay curves to the IH function, one can quantify the energy transfer efficiency through Q and infer the dominant multipolar interaction mechanism through the best-fit exponent s. Some considerations of the IH model include the assumption of static interactions; phonon-assisted energy transfer is not considered. Furthermore, it is possible to distinguish between radiative and non-radiative decay without supporting spectroscopic data.

4.4.1 Influence of Multipolar Interactions

Figure 4.4 presents a simulation of the luminescence intensity decay behavior in Er^{3+}-doped glasses based on the Inokuti–Hirayama (IH) model [21]. The decay curves consider the normalized luminescence intensity, an intrinsic lifetime $\tau_0 = 1.0$ ms and factor $Q = 0.5$, which is typical for Er^{3+} emissions in the infrared region in the transition $^4I_{13/2} \rightarrow {}^4I_{15/2}$ and different values of the interaction parameter s, which corresponds to the type of multipolar interaction between donor and acceptor ions: dipole–dipole ($s = 6$), dipole–quadrupole ($s = 8$), and quadrupole–quadrupole ($s = 10$).

Figure 4.4a shows the evolution of the normalized luminescence intensity $I(t)/I_0$ as a function of the time on logarithmic scale. All curves present non-exponential decay, and the decay becomes increasingly slower with the increment of s values, indicating that long-range energy transfer becomes less efficient as the order of multipolar interaction increases. The dipole–dipole interaction exhibits the fastest decay since the dipole–quadrupole and quadrupole–quadrupole curves decay more slowly. These results underline the sensitivity of luminescence decay dynamics to the nature and strength of ion–ion interactions. Figure 4.4b shows the data in the linearized form based on the logarithm transformation, which follows Eq. (4.12). This transformation allows for a direct graphical verification of the dominant interaction mechanism. Each line corresponds to a distinct s value and reveals a clear distinction in the slopes depending on the multipolar order, which can be used to infer the value of s from experimental data [29].

To gain insight into the energy transfer mechanisms in RE-doped glasses, the dependence of the multipolar interaction strength on interionic distance was analyzed using the generalized relation $\propto 1/R^s$, where s corresponds to different types of coupling between the excited donor ion and the surrounding acceptors [29]. Figure 4.5 illustrates how the interaction strength decreases dramatically with increasing radial distance R, with the decay becoming steeper for higher-order interactions (i.e., $s = 8$ and $s = 10$). At short distances, quadrupole-based interactions are stronger, but their influence decreases rapidly as the distance from the excited ion increases. Although all interaction types exhibit distance-dependent decay, the rate is more pronounced

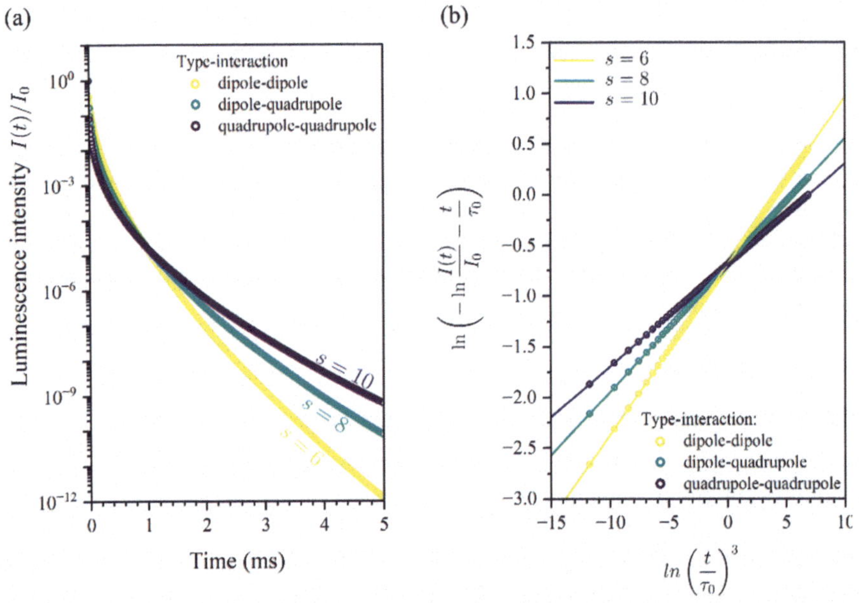

Fig. 4.4 Simulated luminescence decay curves using the Inokuti–Hirayama model for different multipolar interactions. **a** Normalized intensity $I(t)/I_0$ as a function of time for $s = 6$, 8, and 10. **b** Linearized form of the IH model to identify the interaction mechanism from the slope

for larger s values, with the $s = 10$ curve falling off significantly faster than the $s = 6$ case. This trend highlights that, in glass matrices where REIs are randomly distributed, the dominant energy transfer mechanism is strongly influenced by the dopant concentration and spatial configuration. Such graphical representations are useful for interpreting luminescence decay profiles and emphasize how higher-order multipolar interactions are highly sensitive to interionic separation, which directly affects processes such as non-radiative quenching in RE-doped systems.

So far, decay time measurements represent a powerful tool for investigating both radiative and NR processes associated with the energy levels of rare-earth ions. By enabling the analysis of excited-state relaxation dynamics, these measurements provide fundamental insights into the intrinsic properties of electronic transitions, as well as into ion–ion interactions and the influence of the host environment.

Single-exponential decay behavior, typically observed in low-concentration samples and in the absence of significant energy transfer mechanisms, allows direct access to the radiative lifetime. Under such ideal conditions, the decay follows a simple exponential function and enables reliable comparison with theoretical predictions based on Einstein's coefficients. However, real-world systems of technological interest often deviate from this ideal due to the presence of concentration-dependent interactions and competing de-excitation pathways.

Non-exponential decay profiles—arising in highly doped materials, disordered hosts, or systems involving cooperative sensitization and up-conversion—reflect the

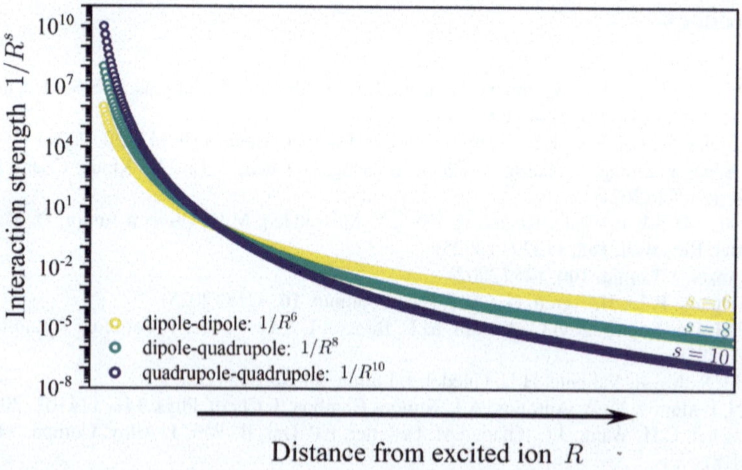

Fig. 4.5 Type intensities as a function of the separation distance between the dipole charges

complexity of the underlying photophysical processes. In such cases, the luminescence decay can no longer be described by a single lifetime constant but instead requires more sophisticated analytical models. Multi-exponential fits, rise-and-decay functions, or stretched exponential functions offer valuable frameworks for capturing the temporal evolution of luminescence and for revealing key aspects of the system, such as energy migration, cross-relaxation, and structural heterogeneity.

The examples presented throughout this chapter highlight how decay analysis can serve as a diagnostic tool for characterizing the efficiency and mechanisms of energy transfer in rare-earth-doped materials. Rather than being a limitation, non-exponential decay behavior provides rich and nuanced information on local environments, quantum efficiency, and the interaction range between dopant ions. These insights are crucial for optimizing luminescent materials, improving device performance, and enabling the rational design of functional photonic platforms.

Finally, the IH model provides a powerful tool for analyzing non-exponential luminescence decay in glasses and the ability to extract the nature of the ion–ion interaction (using s) and the interaction strength (using Q). Glasses doped with ions such as Er^{3+}, Pr^{3+}, and Tm^{3+} in the NIR region benefit from this type of analysis to optimize emission efficiency and energy transfer rates.

In summary, time-resolved luminescence studies, combined with appropriate physical models, constitute an essential approach in the spectroscopic characterization of RE-based systems. They play a central role in advancing emerging technologies in lighting, optical communication, bio-imaging, and energy conversion.

References

1. T.A. Lodi, J.F.M. Dos Santos, G. Galleani, L.G. Jacobsohn, T. Catunda, A.S.S. De Camargo, J. Alloy. Compd. **904**, 164016 (2022)
2. H. Liang, W. Lei, S. Liu, P. Zhang, Z. Luo, A. Lu, Opt. Mater. **119**, 111320 (2021)
3. H. Yang, Y. Zhang, Y. Zhang, Y. Zhao, X. Liang, G. Chen, Y. Liu, W. Xiang, Chem. Eng. J. **414**, 128754 (2021)
4. L.O.E. Da Silva, V.A.G. Rivera, R. Falci, Y. Messaddeq, M.D. Oliveira Junior, D. Manzani, Mater. Res. Bull. **185**, 113291 (2025)
5. F. Auzel, J. Lumin. **100**, 125 (2002)
6. D. Huang, F. Li, H. Ågren, G. Chen, Nat. Commun. **16**, 4218 (2025)
7. J.F.M. Dos Santos, N.G.C. Astrath, M.L. Baesso, L.A.O. Nunes, T. Catunda, J. Lumin. **202**, 363 (2018)
8. G.M. Salley, R. Valiente, H.U. Guedel, J. Lumin. **94–95**, 305 (2001)
9. E.N. Bodunov, Yu.A. Antonov, A.L. Simões Gamboa, J. Chem. Phys. **146**, 114102 (2017)
10. W. Zhai, C.H. Wang, J.C. Qiao, J.M. Pelletier, F.P. Dai, B. Wei, J. Alloy. Compd. **742**, 536 (2018)
11. J.H. Campbell, T.I. Suratwala, J. Non-Cryst. Solids **263–264**, 318 (2000)
12. A.D. Sontakke, K. Biswas, A.K. Mandal, K. Annapurna, Appl. Phys. B **101**, 235 (2010)
13. J. Drabik, R. Kowalski, L. Marciniak, Sci. Rep. **10**, 11190 (2020)
14. S. Mei, J. Zhou, H. Sun, Y. Cai, L. Sun, D. Jin, C. Yan, Adv. Sci. **8**, 2003325 (2021)
15. Z. Wang, A. Meijerink, J. Phys. Chem. C **122**, 26298 (2018)
16. Z. Sui, S. Qian, L. Niu, P. Hu, Z. Hua, X. Zheng, X. Sun, G. Tang, H. Cai, D. Yang, W. Li, M. Zhang, J. Han, J. Ren, Innov. **6**, 100878 (2025)
17. P.M. Tan, N.X. Ca, N.T. Hien, H.T. Van, P.V. Do, L.D. Thanh, V.H. Yen, V.P. Tuyen, Y. Peng, P.T. Tho, Phys. Chem. Chem. Phys. **22**, 6266 (2020)
18. S.S. Nanda, P. Nayak, S.K. Mandal, D. Jana, U.K. Goutam, S. Dash, J. Mol. Struct. **1266**, 133446 (2022)
19. Th., Förster, Ann. Phys. **437**, 55 (1948)
20. D.L. Dexter, J. Chem. Phys. **21**, 836 (1953)
21. M. Inokuti, F. Hirayama, J. Chem. Phys. **43**, 1978 (1965)
22. S. Jackson, *Mid-Infrared Fiber Photonics* (Elsevier, 2022), pp. 305–331
23. L. Giordano, D.L. Fritzen, J.M. Gonçalves, L.C.V. Rodrigues, J.M.A. Caiut, *Modern Luminescence from Fundamental Concepts to Materials and Applications*, vol. 2 (Elsevier, 2025), pp. 101–132
24. A. Chiappini, L. Zur, F. Enrichi, B. Boulard, A. Lukowiak, G.C. Righini, M. Ferrari, *Solar Cells and Light Management* (Elsevier, 2020), pp. 391–414
25. D.D. Ramteke, H.C. Swart, *Upconversion Nanocrystals for Sustainable Technology* (Elsevier, 2025), pp. 133–178
26. J.F.M. Dos Santos, V.S. Zanuto, A.C.C. Soares, E. Savi, L.A.O. Nunes, M.L. Baesso, T. Catunda, J. Lumin. **240**, 118430 (2021)
27. V. Sangwan, M. Jayasimhadri, D. Haranath, J. Mol. Struct. **1322**, 140577 (2025)
28. M. Kuwik, C.K. Jayasankar, W.A. Pisarski, J. Pisarska, J. Alloy. Compd. **842**, 155801 (2020)
29. S.D. Jackson, Opt. Mater. **145**, 114405 (2023)

Chapter 5
Optical Thermometry

The development of instruments for measuring physical and chemical quantities with accuracy and precision remains a huge challenge for researchers. In particular, the measurement of temperature is relevant in a wide variety of fields, e.g., industry, environment monitoring, chemical reactions, and biological processes, among others. In this framework, optical sensors based on glass offer a range of advantages over conventional thermometers, as optical sensors display immunity to electromagnetic interference, the ability to measure in hard-to-reach places, and even remote monitoring due to their wide sampling range along their length, as in the case of optical fibers. Furthermore, optical thermometers exhibit fast response, high thermal and spatial resolution, accuracy, and thermal sensitivity [1–3]. The use of REI-doped glasses for optical sensing is of great interest because REI exhibit spectral responses under temperature changes, which includes the line-shape [4], lifetime [5], wavelength shifting [6], and fluorescence intensity ratio (FIR) [7].

This chapter focuses on the FIR technique, which is based on the band intensity ratio between two thermally coupled energy levels, typically separated by an energy gap of 200–2000 cm^{-1} [8]. Furthermore, the analysis was extended to non-thermally coupled energy levels, providing additional working options. Crucially, the FIR technique is immune to fluctuations in the power of the excitation source, external electromagnetic interference, and energy losses [9]. This is because, for instance, fluctuations in the excitation intensity increase or decrease the collected intensity, giving errors in the temperature calculation. Since such fluctuations affect all bands equally, the ratio remains unchanged. Therefore, the FIR method is perfectly valid for optical thermometry in REI-doped glasses.

G. Lozano C. et al., *A Brief Guide to Optical Spectroscopy of Trivalent Rare Earth Ions in Glasses*, Springer Series in Optical Sciences 247,
https://doi.org/10.1007/978-3-032-18501-3_5

5.1 Fluorescence Intensity Ratio

When the temperature of a REI-doped glass system increases, the electrons of the lower energy level of the ion rise to the higher energy level because the electrons absorb thermal energy in the form of phonons, which has direct consequences on the emission spectrum. In other words, the emission of the lower energy level decreases while that of the higher energy level increases. The emission intensity is in function of the population of the excited i th level N_i, the transition probability A_i, and the emission energy ω_i as $I_i = N_i A_i \hbar\omega_i$ [9]. Since the population of the energy levels obeys the Maxwell–Boltzmann distribution, the FIR between two levels with energies E_1 and E_2 ($E_1 < E_2$) in thermal equilibrium at a temperature T is given by

$$FIR = \frac{I_2}{I_1} = \frac{N_2 A_2 \hbar\omega_2}{N_1 A_1 \hbar\omega_1} = A\exp\left(\frac{-\Delta E}{k_B T}\right) \tag{5.1}$$

where $\Delta E = E_2 - E_1$ and k_B is Boltzmann's constant (0.695 cm^{-1} K^{-1}). Therefore, measuring the band area of each thermally coupled energy level, we can plot a calibration curve of the FIR at a given temperature to obtain by fitting the parameters A and ΔE. This is the regular process for optical thermometry in REI-doped glasses, and several measurements should be performed for accurate results. In this manner, the repeatability of the sensor can be tested in cooling and heating cycles. The performance of a temperature sensor must be evaluated by two sensitivity parameters: (i) the absolute sensitivity S_{abs}, which is the rate of change of the FIR with respect to the temperature and (ii) the relative sensitivity S_{rel}, which is employed to compare the sensitivity of different optical thermometers and thermometry measurement techniques [10]. These sensitivities are calculated from Eq. (5.1) as follows:

$$S_{abs} = \frac{d\mathrm{FIR}}{dT} = A\exp\left(\frac{-\Delta E}{k_B T}\right)\frac{\Delta E}{k_B T^2} \tag{5.2}$$

$$S_{rel} = \frac{1}{\mathrm{FIR}}\frac{d\mathrm{FIR}}{dT} = \frac{\Delta E}{k_B T^2} \tag{5.3}$$

It is relevant to highlight that S_{rel} values larger than $1\%K^{-1}$ are recommended for practical purposes [11]. Furthermore, the performance of optical thermometers is assessed with temperature resolution and the repeatability, denoted as δT and R, respectively. The repeatability is employed to evaluate the precision during the replicate measurements, whereas the temperature resolution determinates the accuracy of the thermal sensor [7, 12]. Both are obtained with the following expressions:

$$\delta T = \frac{1}{S_{rel}}\frac{\delta\mathrm{FIR}}{\mathrm{FIR}} \tag{5.4}$$

$$R = \left(1 - \frac{\mathrm{Max}\{\overline{\mathrm{FIR}} - \mathrm{FIR}_j\}}{\overline{\mathrm{FIR}}}\right) \times 100\% \tag{5.5}$$

Here, $\delta\mathrm{FIR}/\mathrm{FIR}$ is the relative standard deviation of the FIR and $\overline{\mathrm{FIR}}$ is the means of a set of FIR_j values in a specific number of cycles. It is relevant to highlight that R values greater than 90%. Usually, the FIR is calculated by direct integration of each involved band. Nevertheless, sometimes we deal with overlapped bands, and deconvolutions should be performed, and the areas of the obtained bands should be summed. In the following section, we shall do the calculations with both direct integration and sum of the deconvolved bands to corroborate if we find significant differences.

5.2 Experimental Calculation of the FIR

5.2.1 Experimental Setup

For the experimental part, a tellurite glass doped with Yb^{3+} and Er^{3+} will be used, where the compounds used for the glass fabrication are TeO_2, ZnO, and GeO_2, whereas for the doping the compounds are 3 and 0.25 mol% of Yb_2O_3 Er_2O_3, respectively. Since the Yb^{3+} acts as a synthesizer, the glasses will be pumped with a 980 nm diode laser to obtain an efficient UC process of the Er^{3+}. For optical thermometry, it is important to have a cryostat to keep the samples in thermal equilibrium in a controlled manner. In Fig. 5.1, it can be seen an example of an experimental setup for thermal sensing, where the sample is within the cryostat under vacuum, and a coolant is used to reduce the temperature. A thermocouple is employed to measure the temperature of the system and a thermoregulator to control the temperature. Once the system reaches the thermal equilibrium, the measurements are conducted for each temperature in both cooling and heating cycles. To calculate the repeatability, five cycles were carried out.

5.2.2 Spectrum Analysis and FIR Calculation

Figure 5.2 illustrates the normalized emission spectrum at room temperature (298 K) of the Er^{3+} in the range of 500–575 nm, where the bands centered at 524 and 46 nm correspond to the $^2H_{11/2} \rightarrow {}^4I_{15/2}$ and $^4S_{3/2} \rightarrow {}^4I_{15/2}$ transitions. The baseline of the spectrum was removed for the calculations. In Fig. 5.2a since these emission bands overlap, it is considered that the trough, centered at 535 nm, divides both bands. The area of each one is denoted as I_H and I_S and was calculated by direct integration from 500 to 535 nm and from 535 to 575 nm, respectively. For the deconvolution shown in Fig. 5.2b, the spectrum was fitted with multiple Gaussian components via the Levenberg–Marquardt algorithm. The spectrum was deconvolved with at least two Gaussian peaks for the $^4S_{3/2} \rightarrow {}^4I_{15/2}$ band (centered at 542 and 552 nm), and three peaks for the $^2H_{11/2} \rightarrow {}^4I_{15/2}$ band (centered at 518, 523, and 530 nm). Here, the

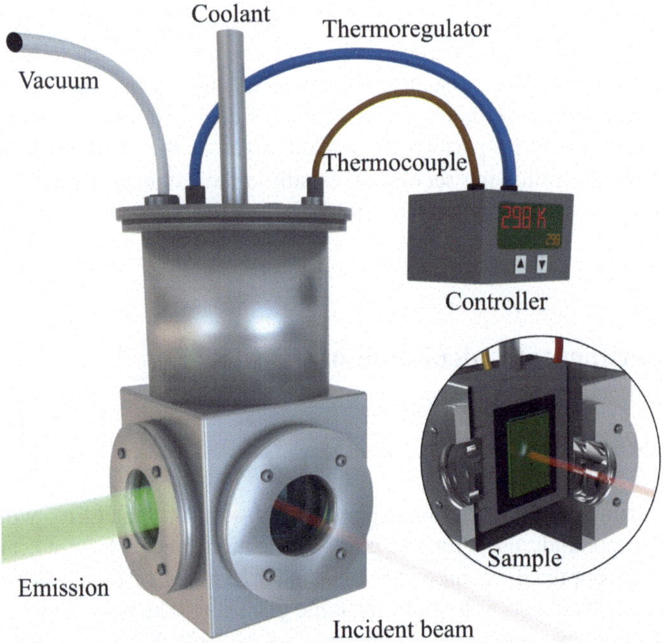

Fig. 5.1 Experimental setup for conventional optical thermometry. The sample is inside the cryostat and in thermal contact with the coolant

area of each Gaussian peak was summed to calculate the total area of each transition. In this manner, the FIR, i.e., I_H/I_S, is calculated for each method, where the FIR for direct integration results in 0.67, whereas by deconvolution is 0.68, which indicates a difference of approximately 1.5%.

Nevertheless, even though deconvolutions give precise band area calculations, one inconvenience is when we deal with low intensities, and the fitting is not accurate enough. Lack of signals appear when we work in both extremely low and high temperatures because under heating, the state with higher energy is populated through NR decays, whereas the lower energy state decreases its population. Hence, it is possible to observe that there are not enough emissions to be detected. Considering these reasons, the temperature range is from 148 to 348 K, and the normalized UC emission spectra under 980 nm excitation are displayed in Fig. 5.3a. It can be observed that with the temperature rising, the $^2H_{11/2} \rightarrow {}^4I_{15/2}$ band increases, conversely, the $^4S_{3/2} \rightarrow {}^4I_{15/2}$ band decreases as expected.

At this point, we have a set of spectra to perform the FIR calculations to fit Eq. (5.1), and Fig. 5.3b and c shows the FIR as a function of the temperature for direct integration and deconvolutions, respectively, with $R^2 > 0.99$. Besides, it is displayed the fitted parameters with their standard errors, where the ΔE values differ about 3%. The ΔE values are lower than reported elsewhere since the energy gaps between the studied thermally coupled levels observed in absorption spectra are about

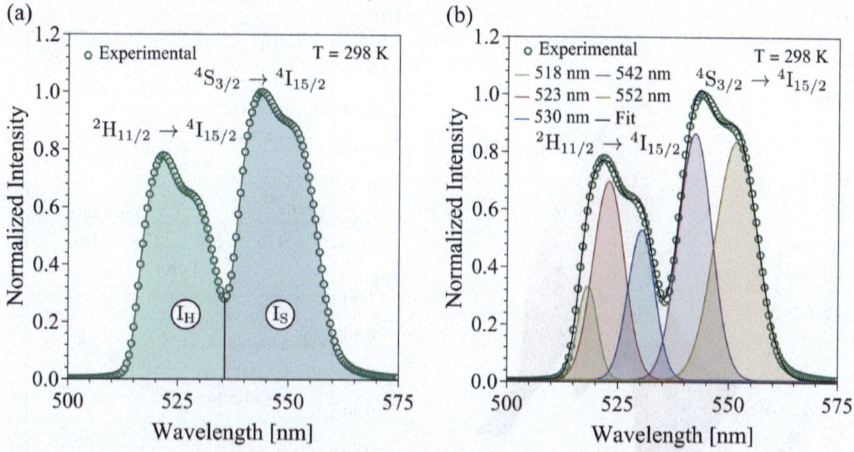

Fig. 5.2 **a** $^2H_{11/2} \rightarrow {}^4I_{15/2}$ and $^4S_{3/2} \rightarrow {}^4I_{15/2}$ emission bands at room temperature, where I_H and I_S denote the area of each band. **b** Deconvolution of the emission bands $^2H_{11/2} \rightarrow {}^4I_{15/2}$ and $^4S_{3/2} \rightarrow {}^4I_{15/2}$ at room temperature ($R^2 > 0.99$)

800 cm^{-1} [13, 14]. The reason for this difference involves transitions between Stark levels of the $^2H_{11/2} \rightarrow {}^4I_{15/2}$ and $^4S_{3/2} \rightarrow {}^4I_{15/2}$ bands, besides the pumping power. For more details about the last statements, the reader is referred to Ref. [7].

Once ΔE and A values are computed, the thermal sensitivities are obtained using both Eqs. (5.2) and (5.3) in an extended temperature range. The plots of both sensitivities can be seen in Fig. 5.3d and e for direct integration and deconvolutions, respectively, where is also highlighted the experimental temperature range used in the measurements (shaded area under the curves). For optical thermometry, it is usual to report the maximum values of S_{abs}, which can be calculated at a specific temperature using the first derivative criteria, i.e., S_{abs} is maximum at $T = \Delta T/2k_B$. Herein, for direct integration and for deconvolutions the maximum values of S_{abs} are 0.43 and 0.46$\%K^{-1}$ at 298 and 307 K, respectively. Regarding S_{rel}, it is common to calculate its value at room temperature (298 K). In this case, the values are 0.67 and 0.69$\%K^{-1}$ for the calculation based on direct integration and deconvolutions, respectively. The calculation of the maximum values of the sensitivities is crucial to compare the performance of the proposed glasses for optical thermometry since the higher the sensitivity value, the higher the accuracy. However, the calculated values in this system are lower than reported elsewhere in other tellurite glasses doped with Er^{3+}-Yb^{3+} [13, 15–17]. Different experimental details can explain the low values for the sensitivities reported here, e.g., larger power values of the excitation source may lead competition between thermalization and other NR processes, the number of experimental data points influences on the calculation of the FIR fitting, and so forth. In a previous study of the glasses used for this chapter, larger values of sensitivities were obtained under different excitation wavelengths, a lower excitation power, and a wider temperature working range [7].

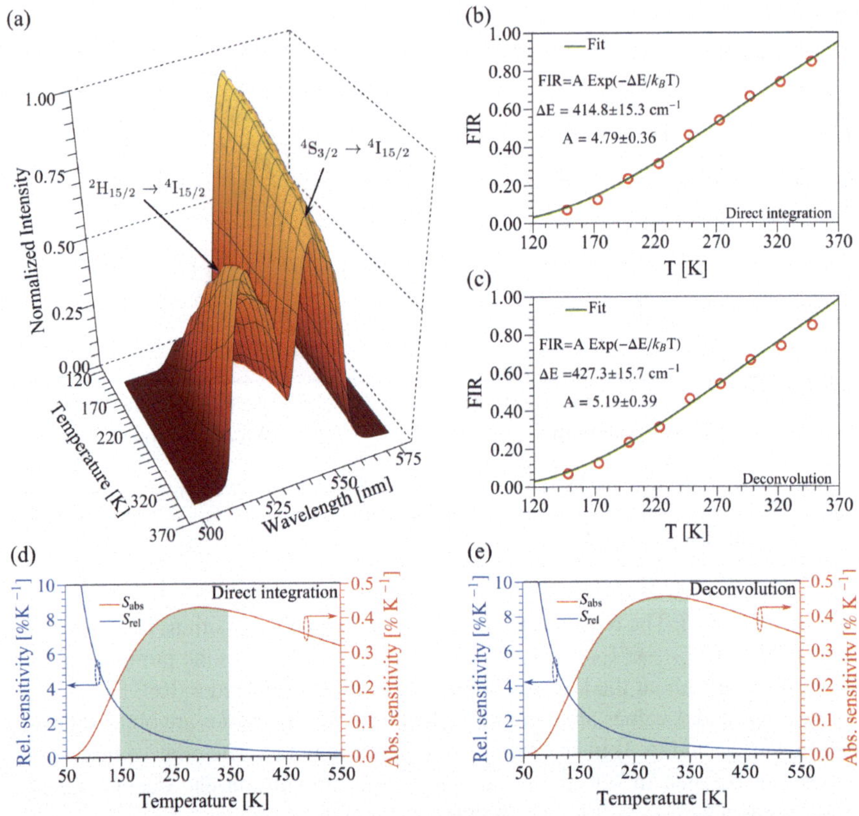

Fig. 5.3 **a** UC emission spectra upon 980 nm excitation at different temperatures. The FIR as a function of the temperature and fitted parameters ($R^2 > 0.99$) using **b** direct integration and **c** deconvolutions. Absolute and relative sensitivities curve as a function of the temperature for **d** direct integration and **e** deconvolutions, where the shaded area indicates the temperature range of the measurements

The FIR was calculated for under the cooling and heating cycles, and their values at both temperature extremes, i.e., 148 and 348 K, are plotted in Fig. 5.4a. The advantage of using optical thermometry is that it offers precise results and large repeatability, as can be seen for the direct integration method, whereas for the deconvolution method, slight deviations are observed. Concerning thermal resolution, it was calculated employing Eq. (5.4) for both methods at a given temperature. In this case, for direct integration, δT was calculated to be 0.30 and 0.77 K at 148 and 348 K, respectively, whereas for deconvolutions was obtained 0.99 and 1.5 K at 148 and 348 K, respectively. Moreover, the repeatability can be computed using the reported FIR values from Fig. 5.4a, where the R values are larger than 99% for direct integration method, while for deconvolutions are 95 and 98% at 148 and 348 K, respectively. The direct integration method offers better repeatability and higher accuracy for temperature sensing, and a straightforward way to calculate the

FIR compared to the deconvolution method. In addition, we can see a comparison between the measured temperature, which is given by the controller (see Fig. 5.1), and the calculated temperature, given by the FIR and the fit parameters, as seen in Fig. 5.4b. In this figure, a continuous line representing the equation $y = x$ is included as the calculated and measured values are expected to coincide ideally. The calculated temperatures exhibit a good agreement with the experimental measurements, remaining reasonably close to the identity line. We can calculate the root-mean-square (RMS) error as $\text{RMS} = \sqrt{(T_m - T_{cal})^2/N}$, where T_m and T_{cal} are the measured and calculated temperatures, respectively, and N is the number of temperature values. The RMSs are 4.89 and 5.31 K for the temperatures calculated by direct integration and by deconvolutions, respectively. Given the temperature range considered (approximately 148–348 K), these deviations represent less than 3% of the full scale, indicating a reasonably good agreement with experimental data.

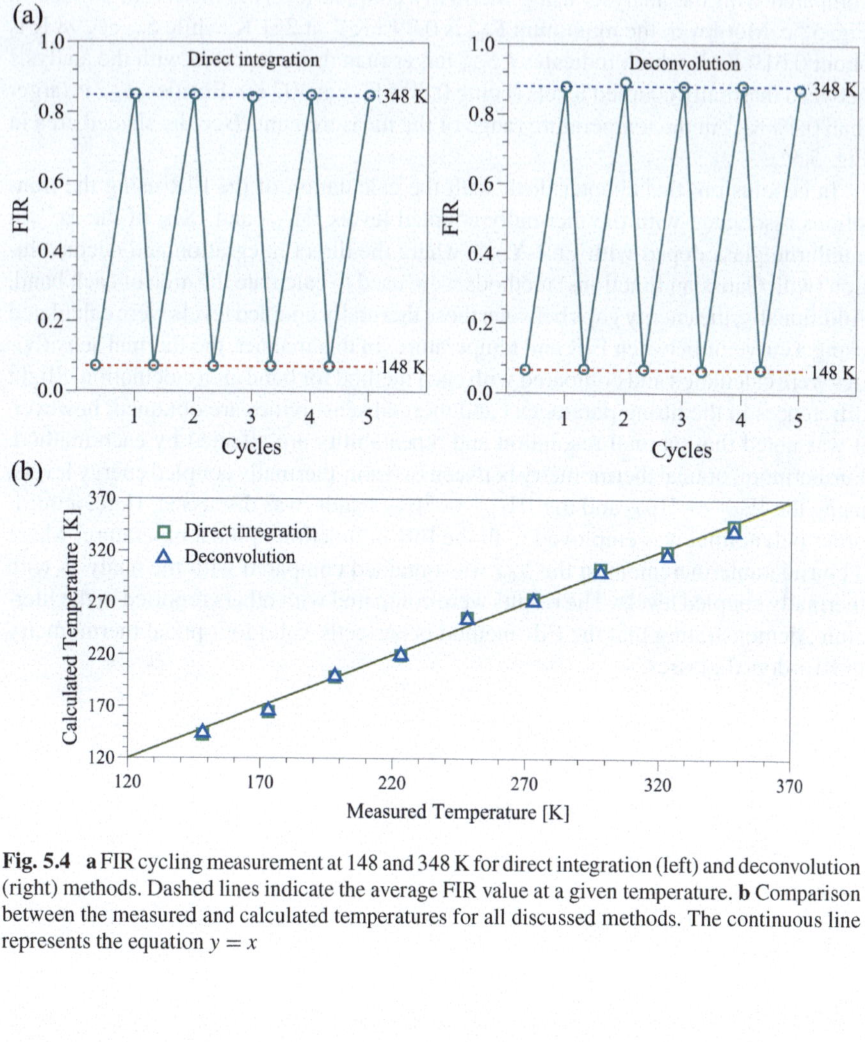

Fig. 5.4 **a** FIR cycling measurement at 148 and 348 K for direct integration (left) and deconvolution (right) methods. Dashed lines indicate the average FIR value at a given temperature. **b** Comparison between the measured and calculated temperatures for all discussed methods. The continuous line represents the equation $y = x$

Sometimes, non-thermally coupled levels are employed for optical thermometry, and these levels do not follow the same approach because of the large energy gap differences, and the sharing of the population between the levels is minimal. For this reason, it is standard to use the polynomial fit of the FIR [18, 19]. Figure 5.5a illustrates the emission spectra upon 980 nm excitation of the studied glass, where the FIR of two non-thermally coupled levels shall be between the $^4F_{9/2} \rightarrow {}^4I_{15/2}$ and the $^2H_{11/2} \rightarrow {}^4I_{15/2}$ (centered at 656 nm) bands. Like the $^2H_{11/2} \rightarrow {}^4I_{15/2}$ emission band, the $^4F_{9/2} \rightarrow {}^4I_{15/2}$ band decreases with temperature. The FIR (direct integration) in function of the temperature is depicted in Fig. 5.5b, and the data points were fitted with a third-order polynomial function $\text{FIR} = a_0 + a_1T + a_2T^2 + a_3T^3$ ($R^2 > 0.99$). The use of a polynomial simplifies the calculation of the S_{abs}; however, when computing the S_{rel}, we can deal with zeroes in the denominator due to normalization. Hence, special attention should be given to the range of temperatures we are analyzing. In our case, from 120 to 370 K, there are no discontinuities, and a similar behavior compared with the analysis using thermally coupled levels is observed, as seen in Fig. 5.5c. Moreover, the maximum S_{abs} is 0.89%K^{-1} at 251 K, while S_{rel} at 298 K is about 0.61%K^{-1}, which indicates a S_{abs} larger than those reported with the analysis between thermally coupled levels (being 0.80%K^{-1} at 307 K). Besides, S_{abs} is larger than 0.4%K^{-1} in the temperature range of the measurements (see the shaded area in Fig. 5.5c).

In conclusion, this chapter dealt with the calculation of the FIR using the transitions associated with the thermally coupled levels $^2H_{11/2}$ and $^4S_{3/2}$ of the Er^{3+} in a tellurite glass doped with Er^{3+}-Yb^{3+}, where the direct integration and deconvolution (with Gaussian functions) methods were used to calculate the area of each band. Additionally, the energy gaps between these thermally coupled levels were calculated using a curve fit between FIR and temperature. In this manner, the thermal sensitivities were calculated and compared with each method for band area calculation. Slight differences in the fitting parameters and thermal sensitivities are obtained; however, it was noted that thermal resolution and repeatability are affected by each method. Furthermore, optical thermometry between two non-thermally coupled energy levels, using the $^4F_{9/2} \rightarrow {}^4I_{15/2}$ and the $^2H_{11/2} \rightarrow {}^4I_{15/2}$ bands, was discussed. Here, a third-order polynomial was employed to fit the FIR in function of the temperature, where a considerable increment in the S_{abs} was obtained compared with the analysis with thermally coupled levels. The results were compared with others reported in the literature, demonstrating that the FIR method is perfectly valid for optical thermometry in REI-doped glasses.

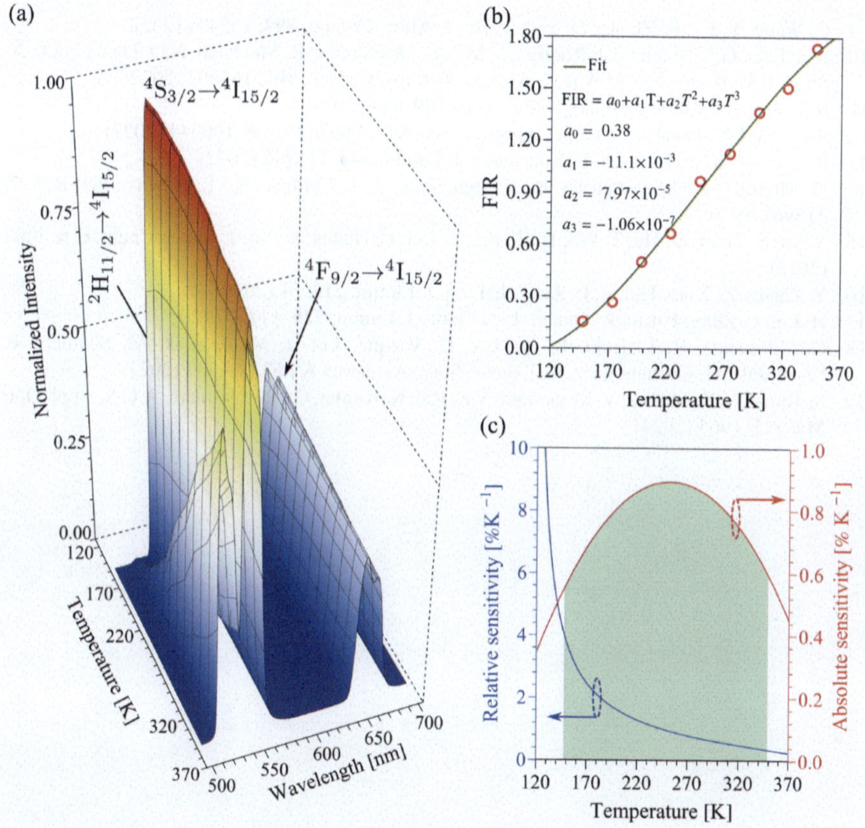

Fig. 5.5 **a** Emission spectra at different temperatures. **b** FIR (by direct integration) as a function of the temperature and the fit with a third-order polynomial. The fit parameters are listed. **c** Relative and absolute sensitivities obtained with the polynomial fit given in **b**. The shaded area indicates the temperature range of the measurements

References

1. J. Zhong, D. Chen, Y. Peng, Y. Lu, X. Chen, X. Li, Z. Ji, J. Alloy. Compd. **763**, 34 (2018)
2. W. Liu, D. Song, Z. Yin, F. Zhang, B. Li, F. Wang, X. Zhang, X. Yan, T. Suzuki, Y. Ohishi, T. Cheng, Opt. Lett., OL **47**, 5297 (2022)
3. D. Wang, T. Zheng, J. Lv, Infrared Phys. Technol. **102**, 102986 (2019)
4. I.C. Pinto, V.A.G. Rivera, R. Falci, V. Fuertes, S. LaRochelle, Y. Messaddeq, J. Lumin. **265**, 120206 (2024)
5. I. Soltani, S. Hraiech, K. Horchani-Naifer, M. Férid, Opt. Mater. **77**, 161 (2018)
6. H. Benrejeb, I.R. Martin, S. Hraiech, K. Soler-Carracedo, Opt. Laser Technol. **149**, 107893 (2022)
7. G.C. Lozano, V. A.G. Rivera, Y. Messaddeq, E. Marega Jr., J. Alloy. Compd. **948**, 169777 (2023)
8. T.O. Sales, C. Jacinto, W.F. Silva, R. Antunes, D.T. Dias, A. Gonçalves, R. El-Mallawany, N.G.C. Astrath, A. Novatski, J. Alloy. Compd. **899**, 163305 (2022)

9. C. Wang, Y. Jin, R. Zhang, Q. Yao, Y. Hu, J. Alloy. Compd. **894**, 162494 (2022)
10. A.S. Laia, G.S. Maciel, J.J. Rodrigues, M.A.C. Dos Santos, R. Machado, N.O. Dantas, A.C.A. Silva, R.B. Rodrigues, M.A.R.C. Alencar, J. Alloy. Compd. **907**, 164402 (2022)
11. R.V. Perrella, P.C.S. de Filho, Dalton Trans. **49**, 911 (2020)
12. S.C. Lal, I.N. Jawahar, S. Ganesanpotti, J. Sci. Adv. Mater. Dev. **8**, 100544 (2023)
13. R. Narro-García, J.J. Leal, E. Rodríguez, J. Lumin. **244**, 118745 (2022)
14. D. Manzani, J.F.S. da Petruci, K. Nigoghossian, A.A. Cardoso, S.J.L. Ribeiro, Sci. Rep. **7**, 41596 (2017)
15. Y. Tu, S. Zhao, D. He, T. Wu, H. Zhang, R. Lei, L. Huang, S. Xu, J. Mater. Chem. C **6**, 7063 (2018)
16. Y. Zhang, Z. Xiao, H. Lei, L. Zeng, J. Tang, J. Lumin. **212**, 61 (2019)
17. H. Lei, L. Zeng, F. Lin, Z. Qin, G. Li, J. Tang, J. Lumin. **229**, 117677 (2021)
18. G.L. Jimenez, R. Lisiecki, B. Starzyk, C. Vazquez-Lopez, M. Lesniak, M. Szumera, P. Szymczak, M. Kochanowicz, D. Dorosz, Sens. Actuators A **357**, 114367 (2023)
19. S. Tomar, N.K. Mishra, V. Kesarwani, V.K. Rai, K. Kumar, C. Shivakumara, A.C.S. Appl, Opt. Mater. **2**, 1965 (2024)

Chapter 6
Multiphonon Processes and Energy Transfer Processes

Multiphonon relaxation mechanisms and energy transfer (ET) processes [1–4] play a critical role in determining the luminescent properties of rare-earth ions (REI)-doped materials [5–7]. These processes directly influence emission efficiency, color, and intensity, and are fundamental to understanding the dynamics of excited-state deactivation.

Multiphonon relaxation [8] constitutes a non-radiative pathway through which the energy of excited states is dissipated into the host lattice in the form of vibrational (phonon) energy [9]. This mechanism becomes particularly efficient when the energy gap between adjacent electronic levels is small and the host matrix exhibits high phonon energy. Consequently, multiphonon processes often act as quenching mechanisms, reducing excited-state lifetimes and lowering quantum efficiency [10]. However, this characteristic does not necessarily represent a drawback. In contrast, it can be strategically exploited to facilitate ET processes that would otherwise be inefficient or unlikely in the absence of phonon-assisted relaxation pathways [11]

Energy transfer interactions are also central to the photophysical behavior of REIs [4, 12, 13]. For instance, energy migration between the sensitizer and activator ions can either enhance or suppress specific emissions, depending on the spatial distribution of the ions and the alignment of their energy levels. This is particularly relevant in systems designed for up-conversion, quantum-cutting, or down-conversion applications, where ET processes critically influence the resulting photonic output.

This chapter provides a detailed discussion of the interplay between energy transfer and multiphonon relaxation in REI-doped materials, emphasizing their combined impact on luminescence decay dynamics and overall optical performance. By examining how these mechanisms depend on dopant concentration, energy level structure, and host matrix characteristics, we outline a framework for designing more efficient phosphors, lasers, and luminescent devices.

G. Lozano C. et al., *A Brief Guide to Optical Spectroscopy of Trivalent Rare Earth Ions in Glasses*, Springer Series in Optical Sciences 247,
https://doi.org/10.1007/978-3-032-18501-3_6

6.1 Multiphonon Processes

Solid matrices doped with rare-earth ions can undergo electronic transitions mediated by interactions with the vibrational modes of the host material. In both crystalline and amorphous (glass) hosts, such transitions are typically associated with the emission or absorption of phonons—quantized units of lattice vibrations [14]. Before proceeding with the discussion, some fundamental phonon-related concepts are briefly introduced to aid in understanding:

(i) A phonon represents a discrete quantum of vibrational energy and corresponds to the elementary oscillatory motion of atoms around their equilibrium positions [14].
(ii) In a crystalline solid, a normal mode is defined as a collective oscillation in which all atoms vibrate coherently at the same frequency. These normal modes are mutually independent.
(iii) The overall lattice vibrations can be described as a superposition of several such normal modes.
(iv) The phonon density of states refers to the number of phonon modes that fall within a specific frequency interval. It spans from zero up to the maximum phonon frequency supported by the crystal lattice.

Although glasses lack long-range translational symmetry—resulting in vibration modes without well-defined wave vectors, such excitations are still conventionally referred to as phonons for consistency. When the energy separation between electronic states is small enough to be bridged by the emission or absorption of one or two phonons (e.g., between Stark levels within a multiplet), these transitions occur efficiently and rapidly. As a result, a thermal population of excited sublevels above the ground or metastable states can be established when the energy gaps are on the order of a few kT. In contrast, energy separations between different J-multiplets are typically much larger than kT, and non-radiative transitions between them proceed predominantly in the downward direction via the emission of multiple phonons to conserve energy. When the non-radiative relaxation rate becomes comparable to or exceeds the radiative decay rate, the luminescence efficiency associated with that excited level is significantly reduced.

Non-radiative processes are particularly critical in determining the overall quantum efficiency of luminescent materials, especially for transitions involving excited states with small energy gaps to lower-lying levels. In these cases, the emission of multiple phonons can efficiently depopulate the excited state before photon emission occurs, reducing the effectiveness of the material in applications such as lasers, amplifiers, and phosphors.

The investigation of multiphonon relaxation processes in rare earth ions began in the mid-twentieth century [3] The theoretical framework was initially developed for crystalline hosts by Kiel [2] and later extended by Riseberg and Moos [15], pioneering determination the non-radiative decay rates (W_{NR}) of various Ln^{3+} 4f states in various lattices based on the analysis of excited-state dynamics. Since these transitions occur across energy gaps that are often several times larger than the energy of the highest-energy phonons, their description requires high-order perturbation theory. While high-order electromagnetic processes—such as multiphoton transitions—are typically very weak, multiphonon relaxation can be significant due to the stronger electron–phonon coupling and the much higher phonon density of states, which is typically 11 orders of magnitude greater than that of photons [2]. Because the transition probability decreases rapidly with the number of phonons needed to bridge the energy gap, the dominant contribution to non-radiative decay arises from the highest-energy phonon modes available in the host material.

The strong dependence of vibrational spectra on the host composition makes the non-radiative relaxation rate highly hosted. In contrast, for a given host, levels separated by similar energy gaps tend to exhibit comparable non-radiative decay rates, largely independent of the detailed electronic structure or even the identity of the rare-earth ion involved—except in cases where stringent selection rules apply.

For energy gaps significantly larger than the phonon energy, W_{NR} follows an exponential dependence on the gap, as expressed by the empirical relation [4, 8]:

$$W_{NR} = C[n(T) + 1]^p \exp(-\alpha \Delta E) \tag{6.1}$$

where ΔE is the energy gap, α and C are host-dependent parameters, p is the number of phonons required to bridge that gap, and $n(T)$ is the Bose–Einstein occupation number for the effective phonon mode,]

$$n(T) = \frac{1}{\exp\left(\frac{\hbar\omega}{kT}\right) - 1} \tag{6.2}$$

where ω represents the frequency of the phonons. As temperature decreases, the non-radiative decay state also decreases due to the temperature dependence embedded in the phonon occupation number $n(T)$. In practical applications, the parameters C, α, and p are treated as empirical and are primarily determined by the host material. These parameters are generally considered to be insensitive to the specific rare earth ion or the electronic levels involved. Their values are typically obtained by fitting Eq. (6.1) to experimentally measured non-radiative decay rates across a wide range of energy gaps, using multiple energy levels and rare earth dopants within the same host matrix.

The frequency of phonons serves as one of the keys to assess the probability of non-radiative relaxation of REI^{3+}. Van Dijk et al. [16] provided a comprehensive analysis of p-dependent non-radiative rates for various Ln^{3+} ions across a wide range of host materials. Their findings revealed a convergence point closer to $p = 2$, with typical relaxation rates near 10^7 s^{-1}.

Since the energy spacing between the 4f states of REI^{3+} are only marginally affected by the surrounding matrix, the relaxation probability is governed primarily by variations in the host's phonon spectrum. In most cases, the phonons that dominate the non-radiative relaxation process correspond to the highest-frequency vibrations

($\hbar\omega_{max}$) of the host, as these provide the lowest-order multiphonon processes and typically exhibit high densities of states.

Indeed, a decrease in the energy gap leads to an exponential increase in W_{NR}, underscoring the dominance of multiphonon relaxation processes when the separation between RE^{3+} 4f states is small. In such systems, a new dependence of the multiphonon relaxation rate on p involved has been established, where p is calculated by dividing the energy gap by the $\hbar\omega_{max}$ of the host lattice [16]. Experimental findings have demonstrated that W_{NR} becomes the prevailing mechanism for electronic de-excitation when the energy gap corresponds to fewer than five phonons [1]. This "five-phonon" threshold has since been adopted as a practical rule of thumb to estimate the relaxation probability of RE^{3+} 4f excited states. For instance, the $^4I_{11/2} \rightarrow {}^4I_{13/2}$ transition in Er^{3+} has an energy gap of ~3700 cm^{-1}. In hosts with high-energy phonons like phosphate glasses (phonon energy ~1200 cm^{-1}) [17], this gap can be bridged by just three phonons, leading to a high non-radiative decay rate and a substantial reduction in the luminescence efficiency at 1.5 μm.

Similarly, the $^3H_4 \rightarrow {}^3F_4$ transition in Tm^{3+}, with an energy gap of approximately 5000 cm^{-1}, is strongly quenched in high-phonon-energy hosts such as silica [18] (phonon energy ~1100–1200 cm^{-1}), since the gap can be bridged by fewer than five phonons. This efficient multiphonon relaxation limits the up-conversion emission in the blue region (~480 nm), commonly observed in Tm^{3+}-doped systems.

Another example is the $^4F_{3/2} \rightarrow {}^4I_{11/2}$ transition in Nd^{3+}, which has an energy separation of ~2000 cm^{-1}. In oxide-based glasses, this gap corresponds to fewer than two phonons, making the level highly susceptible to non-radiative de-excitation [15] [1]. As a result, the laser performance of Nd^{3+} in such hosts is significantly compromised. In contrast, in low-phonon-energy materials such as fluorides or chalcogenides, the same transition exhibits reduced non-radiative rates, enhancing radiative efficiency.

A further case is observed with Ho^{3+} ions, where the $^5I_6 \rightarrow {}^5I_7$ transition (~2000 cm^{-1}) is prone to multiphonon quenching in oxide matrices, but becomes radiatively allowed in fluoride hosts, supporting mid-infrared emission around 2.9 μm [19]. This illustrates how host-dependent phonon energy plays a critical role in enabling or suppressing specific emission bands.

These examples collectively emphasize the importance of host material selection in the design of efficient luminescent systems, particularly for applications in lasers, optical amplifiers, and emission-based devices.

6.2 Energy Transfer Microparameters

In REI–doped materials, two fundamental energy transfer mechanisms are typically observed: radiative and non-radiative processes. The radiative mechanism involves two sequential stages—emission of radiation by the donor (D) ion, followed by absorption of this radiation by the acceptor (A) ion—as described below:

$$D^* \rightarrow D + h\nu \tag{6.3}$$

$$A + h\nu \rightarrow A^* \tag{6.4}$$

The non-radiative ET process, on the other hand, consists of a single step that simultaneously involves the decay of the D ion and the excitation of the A ion:

$$D^* + A \rightarrow D + A^* \tag{6.5}$$

In the 1920s, Perrin introduced a theoretical model for energy transfer based on resonance interactions between the electric dipoles of donor and acceptor molecules. However, certain assumptions in this model—particularly the premise that identical molecules possess perfectly matched oscillation frequencies—led to significant overestimations of the energy transfer radius, with predicted distances reaching several hundred to thousands of angstroms.

Subsequently, Förster refined Perrin's model using quantum mechanics and perturbation theory, producing energy transfer distances more consistent with experimental observations. While Förster's model accounts exclusively for dipole–dipole interactions, Dexter [20] later extended the theoretical framework to include cases where dipole-mediated energy transfer is forbidden or inefficient, introducing mechanisms based on higher-order multipolar and exchange interactions. These theoretical models are analyzed in the following sections.

6.2.1 Förster–Dexter Theory

Förster [21, 22] described the transfer of energy from a set of donor atoms to a set of acceptor atoms (donor–acceptor interaction) through electric dipole–dipole interaction. The interaction rate W_{DA} is proportional to the absolute square of the dipole–dipole interaction Hamiltonian $\hat{H}_{d-d}$,

$$\widehat{H}_{d-d} = \frac{1}{4\pi\varepsilon_0}\left[\widehat{\mu}_D \bullet \widehat{\mu}_A - 3(\widehat{\mu}_D \bullet \hat{r})(\widehat{\mu}_A \bullet \hat{r})\right]\frac{1}{R^3} \tag{6.6}$$

$$W_{DA} \propto \left|\widehat{H}_{d-d}\right|^2 \tag{6.7}$$

where R is the distance between the dipoles (donor/acceptor), $\hat{r}$ is the unit vector, $\widehat{\mu}_D$ is the donor dipole moment, and $\widehat{\mu}_A$ is the acceptor dipole moment. Generally, the donor and acceptor atoms are of different species; however, there is the possibility of energy transfer to an atom of the same species, i.e., donor–donor interactions. The probability per unit time for dipole–dipole interaction is:

$$W_{DD/DA} = \frac{1}{\tau_{D,rad}}\left(\frac{R_{DD/DA}}{R}\right)^6 = \frac{C_{DD/DA}}{R^6} \tag{6.8}$$

Here $\tau_{D,rad}$ is the radiative lifetime of the donor in the absence of this process, $R_{DD/DA}$ is the Förster radius for donor–donor or donor–acceptor energy transfer, defined as the distance at which the rate of energy transfer by dipole–dipole coupling equals the luminescence decay rate, $C_{DD/DA}$ are the donor–donor and donor–acceptor microparameters, and the exponent 6 denotes the multipolarity of the dipole–dipole interaction. Notably, when the donor–acceptor separation equals the Förster radius ($R_{DD/DA}$ =R), the rate of energy transfer matches the radiative decay rate of the donor in the absence of an acceptor ($W_{DD/DA} = \frac{1}{\tau_{D,rad}}$). As demonstrated in Eq. (6.8), this rate exhibits a strong R^{-6} dependence, which originates from the dipole–dipole interaction scaling with R^{-3} Eq. (6.6). Consequently, the transfer probability, being proportional to the square of this interaction, follows an inverse sixth-power law with distance.

The radius of Förster and the microparameters can be calculated through the cross-section overlap function [23]:

$$C_{DD/DA} = \frac{R^6_{DD/DA}}{\tau_{D,rad}} = \frac{3c}{8\pi^4 n^2} \int \sigma^D_{em}(\lambda)\sigma^{D/A}_{abs}(\lambda)d\lambda \tag{6.9}$$

where c is the speed of light, n is the refractive index of the material, and $\sigma^D_{em}(\lambda)$, are the emission cross-section spectra and $\sigma^{D/A}_{abs}(\lambda)$ is the absorption cross-section spectra, respectively, at wavelength λ. The processes of energy migration and ET are described by transition probabilities equal to W_{DD} and W_{DA}, respectively. The integral presented in Eq. (6.4) is also known as spectral overlap integral, $J(\lambda)$. A higher $J(\lambda)$ value indicates a greater overlap and thus, a higher probability of energy transfer between the donor and acceptor, as presented in Fig. 6.1

Dexter's model [20] extends Förster's theory by incorporating exchange interactions and higher-order multipolar mechanisms, such as dipole–quadrupole and quadrupole–quadrupole interactions, whose energy transfer probabilities scale with distance as R^{-8} and R^{-10}, respectively. These mechanisms become relevant when one or both transition dipoles are forbidden. For example, energy transfer involving triplet states is often dominated by exchange interactions, which are governed by the overlap of the donor and acceptor wavefunctions. Dexter energy transfer—also called Dexter electron transfer—is a fluorescence quenching mechanism in which an excited electron is transferred non-radiatively from a donor molecule to an acceptor molecule. This process requires overlap between the electronic wavefunctions of the donor and acceptor, which restricts it to short-range interactions, typically within 10 Å (1 nm). The energy transfer can occur through either a single-step exchange of the excited state or via two sequential charge transfer steps. The Dexter energy transfer rate constant, W_{Dexter}, is given by

$$W_{Dexter} = KJ(\lambda)e^{[-2R/L]} \tag{6.10}$$

where K is an experimental factor related to the orbital interactions and L is the sum of van der Waals radii of the donor and acceptor. Because K, is a constant that

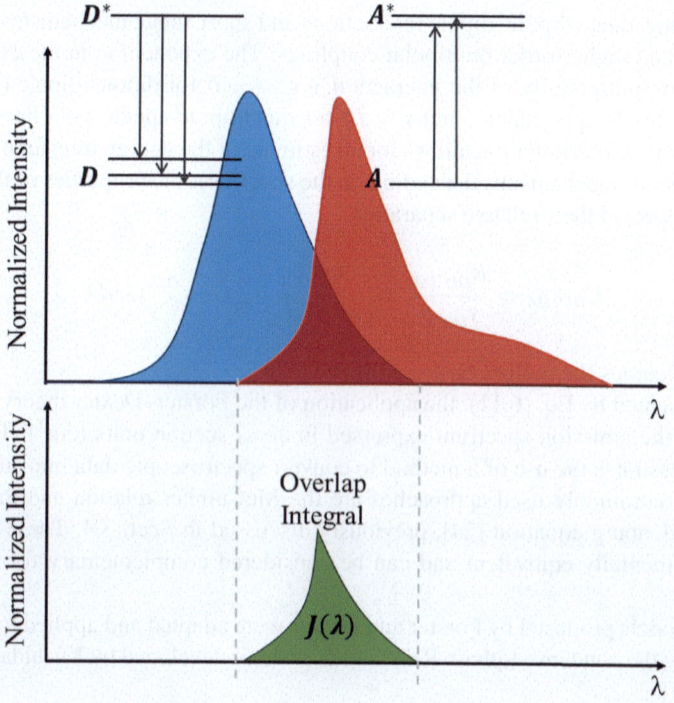

Fig. 6.1 Schematic representation of the integral of spectral overlap, $J(\lambda)$ (green area), between the normalized emission spectrum of the donor (blue curve) and the absorption spectrum of the acceptor (red curve). Vertical arrows indicate electronic transitions: downward arrows correspond to donor emission, while upward arrows represent acceptor excitation

is not related to any spectroscopic data, it is difficult to characterize the exchange mechanism experimentally. Table 6.1 summarizes the main characteristics of the Förster and Dexter energy transfer models.

Based on these mechanisms, the general expression for the energy transfer rate constant for donor–donor or donor–acceptor pairs can be written as Eq. (6.11), commonly referred to as the *Förster–Dexter equation* [13]. This equation accounts

Table 6.1 Förster versus dexter energy transfer

Feature	Förster	Dexter
Type of interaction	Dipole–dipole	Electron exchange (wave-Function overlap)
Distance dependence	$\propto R^{-6}$	$\propto e^{[-2R/L]}$
Typical range	1–10 nm	<1 nm
Spectral overlap required	Yes	Yes
Mechanism	Long-range Coulombic coupling	Short-range orbital overlap

for both long-range dipole–dipole interactions and short-range mechanisms, such as exchange and higher-order multipolar couplings. The exponent m in the expression reflects the multipolarity of the interaction, e.g., $s = 6$ for dipole–dipole (Förster), $s = 8$ for dipole–quadrupole, and $s = 10$ for quadrupole–quadrupole interactions. This generalized formulation allows for an estimate of the energy transfer efficiency for a variety of mechanisms, depending on the spectroscopic properties of the donor and acceptor and their relative separation.

$$C_{DD/DA} = \frac{R^{s}_{DD/DA}}{\tau_{D,rad}} = \frac{3c}{8\pi^4 n^2} \int \sigma^{D}_{em}(\lambda)\sigma^{D/A}_{abs}(\lambda)d\lambda \qquad (6.11)$$

where m denotes the multipolarity of the process.

As indicated by Eq. (6.11), the application of the Förster–Dexter theory requires access to the emission spectrum expressed in cross-section units (cm^2). This typically necessitates the use of a method to convert spectroscopic data into such units. The most commonly used approaches are the McCumber relation and the Füchtbauer–Ladenburg equation [24], previously discussed in Sect. 3.4. These methods are fundamentally equivalent and can be considered complementary for practical purposes.

The models proposed by Förster and Dexter were adapted and applied to systems containing the condensed-phase REI^{3+}, such as those developed by Kushida [12] and Malta [25].

6.2.2 Linking Decay Kinetics to Microscopic Energy Transfer Parameters

In Chap. 4, we analyze that Inokuti and Hirayama developed a theoretical model describing the relationship between energy transfer rate and luminescence decay, which has since been widely adopted—either in its original formulation or through various adaptations. A key assumption of the model is that energy transfer occurs exclusively between donor and acceptor ions, while donor–donor interactions are considered negligible. Burshteîn [26] expanded the Inokuti–Hirayama model by incorporating donor–donor interactions. In the hopping regime of donor–donor energy transfer—considered applicable to rare-earth ions [4, 13]—the excitation can reach a sphere surrounding the acceptor, defined by the Förster radius R_{DA}, in a single discrete step. At the boundary of this sphere, the donor excitation has an equal probability of being quenched by the acceptor or migrating out of the interaction volume. The hopping regime is established when:

$$\frac{C_{DA}}{C_{DD}} \ll N_D^{(s_{DD}-s_{DA})/3} \qquad (6.12)$$

where N_D represents the donor concentration. For dipole–dipole interactions—i.e., when both donor–donor and donor–acceptor interactions exhibit a multipolarity of s_{DD}=s_{DA}=6—the condition for the hopping regime becomes $C_{DA}/C_{DD} \ll 1$. According to Burshteĭn [26], when a donor is excited at t = 0, the probability $f_D(t)$ of finding a donor, which can transfer energy to acceptors that it remains in the excited state decreases over time due to two main processes: the intrinsic radiative decay, characterized by the lifetime τ_D, and impurity quenching, in which the excitation energy is transferred from the donor to an acceptor. This leads to the following expression:

$$f_D(t) = exp\left(-\frac{t}{\tau_D}\right)P(t) \tag{6.13}$$

where $P(t)$ represents a complex multistep process that depends on both the donor and acceptor concentrations, N_D and N_A, respectively. The rate of this process is governed by the migration of excitations among donor sites. The function $P(t)$ satisfies the following rate equation:

$$\frac{dP}{dt} = -k(t)N_A P \tag{6.14}$$

In this context, the term $k(t)N_A$ corresponds to the time-dependent quenching rate. When evaluated within the framework of the hopping regime, Eq. (6.14) leads to an analytical expression that describes the temporal behavior of the luminescence decay.

$$I(t) = I(0)exp\left[-\frac{t}{\tau_D} - \gamma\sqrt{t}\right]exp(-\overline{W}_t t) \tag{6.15}$$

where $\gamma = -\frac{4}{3}\pi^{3/2}N\sqrt{C_{DA}}$ and $\overline{W}_t = \pi\left(\frac{2\pi}{3}\right)^{5/2}\sqrt{C_{DA}C_{DD}N^2}$. $\overline{W}_t$ is an ensemble-average migration-assisted energy transfer rate. Equation (6.14) is the product of Inokuti–Hirayama's static (which was analyzed in exponential term dependent on energy-migration exponential term.

The expression given by Eq. (6.15) highlights that the decay kinetics in the hopping regime is not purely exponential due to the contribution of donor excitation migration prior to the quenching event. To further refine the theoretical treatment of energy transfer in concentrated systems, Yokota and Tanimoto [27] proposed an alternative approach by introducing a time-dependent interaction radius. In their model, the excitation does not immediately experience the full influence of all acceptors within the critical volume, but rather progressively explores the surrounding medium as time evolves.

The Yokota–Tanimoto model is particularly suitable for systems in which energy migration among donors significantly precedes transfer to acceptors—a common scenario in rare-earth-doped materials at moderate to high donor concentrations [27–29]. Unlike the Inokuti–Hirayama model, which assumes a static spatial distribution of acceptors, the Yokota–Tanimoto formulation incorporates the dynamic expansion of the donor's migration sphere over time, leading to a modified expression for the donor decay probability [27, 30]:

$$I(t) = I_0 exp\left[-\frac{t}{\tau_{D,rad}} - \frac{4\pi}{3}N_A\Gamma\left(1-\frac{3}{s}\right)(C_{DA}t)^{3/s}\left(\frac{1+10.87X+15.5X^2}{1+8.74X}\right)^{\frac{s-3}{s-2}}\right] \tag{6.16}$$

where $X = DC_{DA}^{-\frac{2}{s}}(t^{1-\frac{2}{s}})$ and $D = \frac{1}{2}\left(\frac{4\pi N_D}{3}\right)^{4/3}C_{DD}$. The parameters, D and N_D, are the diffusion coefficient and N_D the donor ion concentration, respectively. This model bridges the gap between the purely static regime described by Inokuti–Hirayama and the fully dynamic hopping regime addressed by Burshteîn, offering a more comprehensive framework for interpreting h experimental decay curves, especially when the migration and transfer rates are comparable.

In summary, while the Inokuti–Hirayama model laid the foundation for understanding luminescence decay in dilute systems and Burshteîn extended it by incorporating migration effects explicitly, the Yokota–Tanimoto model adds further nuance by accounting for the time-dependent evolution of the donor's interaction environment. Together, these models provide a robust theoretical toolkit for analyzing energy transfer dynamics in a wide variety of doped luminescent materials.

These theoretical models are particularly useful for determining microscopic parameters related to the energy transfer mechanisms that occur in REI-doped materials. By providing a quantitative description of donor–acceptor and donor–donor interactions, such models allow the estimation of critical parameters such as the Förster radius, energy transfer rates, and quenching efficiencies. Consequently, they contribute significantly to the understanding and prediction of macroscopic properties of the material, including luminescence efficiency, lifetime behavior, and concentration quenching effects. These insights are essential for the design and optimization of photonic materials with customized optical properties.

6.2.2.1 Determination of the Microparameters C_{DD} and C_{DA}: Case Study in Nd^{3+}-Doped Glasses

Let us consider the laser transition $^4F_{3/2} \rightarrow {}^4I_{11/2}$ in Nd^{3+}-doped glasses [17, 23, 31]. At low concentrations, the quenching of the $^4F_{3/2}$ level is primarily governed by two non-radiative processes: (i) energy migration among excited Nd^{3+} ions (donor–donor transfer), and (ii) cross-relaxation between excited and ground-state ions (donor–acceptor transfer), indicate in Fig. 6.2a. The microscopic parameters C_{DD} and C_{DA} quantify the strength of these dipole–dipole interactions.

(i) The C_{DD} microparameter is associated with energy migration between Nd^{3+} ions and involves the transitions $^4F_{3/2} \rightarrow {}^4I_{9/2}/{}^4I_{9/2} \rightarrow {}^4F_{3/2}$, as illustrated by the green dashed arrows. Since the initial and final energy levels involved are identical,

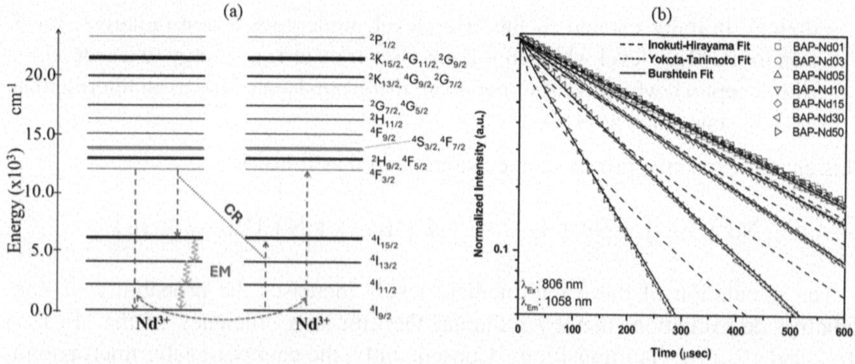

Fig. 6.2 **a** Schematic energy-level diagram of Nd^{3+} ions in doped glasses, illustrating the main non-radiative processes responsible for the depopulation of the excited $^4F_{3/2}$ level: energy migration (EM) between Nd^{3+} ions via resonant transfer (green arrows) and cross-relaxation (CR) with neighboring ground-state ions (blue arrows). These processes correspond to donor–donor and donor–acceptor energy transfer mechanisms, respectively. **b** Semi-logarithmic plot of luminescence decay curves for Nd^{3+}-doped barium–alumino–metaphosphate glasses (BAP-Nd) with varying Nd^{3+} concentrations, following excitation at 806 nm and emission monitoring at 1058 nm ($^4F_{3/2} \rightarrow {}^4I_{11/2}$ transition). The experimental data (symbols) are fitted using three theoretical models: Inokuti–Hirayama (static regime), Yokota–Tanimoto (diffusion-controlled regime), and Burshteîn (hopping regime with donor migration). The comparative fits illustrate the increasing influence of energy migration and cross-relaxation processes with dopant concentration

this energy transfer process is resonant in nature, enabling the application of the Förster–Dexter theory.

As discussed earlier, the calculation of the donor–donor energy transfer C_{DD} microparameter relies on the application of Eq. (6.11), which requires prior knowledge of the emission cross-section $\sigma_{em}(\lambda)$. To obtain this parameter, Sontakke et al. [30] analyzed the luminescence properties of Nd^{3+}-doped barium-alumino-metaphosphate glasses (BAP-Nd) over a wide concentration range, from 0.276 to 13.31 × 10^{20} ions/cm^3, for which the average donor–donor distance R_{DD} was found to decrease from 33.1 Å to 9.1 Å.

Following the procedures detailed in Chap. 3, $\sigma_{em}(\lambda)$ was determined using two complementary approaches [24]: McCumber's reciprocity relation Eq. (3.1) and the Füchtbauer–Ladenburg Eq. (3.2). With the emission cross-sections thus obtained, the authors applied Eq. (6.11) to calculate the microparameter C_{DD} for the $^4F_{3/2} \rightarrow {}^4I_{9/2}$ transition. The resulting values were approximately 3.8 × 10^{-39} cm^6/s (using McCumber's method) and 3.2 × 10^{-39} cm^6/s (using the Füchtbauer–Ladenburg relation), confirming the consistency between both techniques and demonstrating the effectiveness of the approach for quantifying donor–donor dipole–dipole interactions.

(ii) The authors also determined, for this glass system, the microparameter C_{DA}, which is related to the CR process, represented in Fig. 6.2a by the blue arrows. In the CR mechanism, involving two neighboring Nd^{3+} ions, one ion (the donor), initially excited to the $^4F_{3/2}$ level, undergoes a non-radiative transition to the $^4I_{15/2}$ level while simultaneously transferring energy to a second ion (the acceptor), which is promoted from the ground state $^4I_{9/2}$ to an intermediate excited state, such as $^4I_{13/2}$ or $^4I_{15/2}$.

This energy exchange process can be summarized as follows:

$$Nd^3\left(^4F_{3/2}\right) + Nd^3\left(^4I_{9/2}\right) \rightarrow Nd^3\left(^4I_{15/2}\right) + Nd^3\left(^4I_{15/2}/\,^4I_{13/2}\right)$$

The population of these intermediate levels increases the probability of non-radiative de-excitation, thereby reducing the emission efficiency of the $^4F_{3/2} \rightarrow$ $^4I_{11/2}$ and $^4F_{3/2} \rightarrow$ $^4I_{9/2}$ transitions. Consequently, the energy transfer microparameter C_DA, associated with donor–acceptor energy transfer via the CR mechanism, can be estimated through the analysis of donor luminescence decay using different theoretical models, as presented in Sect. 6.2.2.

According to previous studies, energy transfer between Nd^{3+} ions is predominantly governed by dipole–dipole electrostatic interactions (with interaction parameter $s = 6$). Based on this assumption, the donor fluorescence decay spectra were analyzed using theoretical fits derived from the Burshteîn and Yokota–Tanimoto models, as also presented in Fig. 6.2b.

For instance, using the Yokota–Tanimoto model, a sample with intermediate Nd^{3+} concentration (BAP–Nd05) was obtained $C_{DA} \approx 8.57 \times 10^{-42}$ cm^6/s; $R_{DA} \approx 3.78$ Å and, D $\approx 0.12 \times 10^{-11}$ cm^2/s. At higher concentrations, such as in BAP–Nd50, diffusion becomes more significant (D $\approx 2.09 \times 10^{-11}$ cm^2/s), while C_{DA} remains on the order of 4.86×10^{-42} cm^6/s, and $R_0 \approx 3.44$ Å.

Similarly, fits using the Burshteîn model, which accounts for hopping-mediated migration, provided C_DA values ranging from 7.09×10^{-42} cm^6/s (BAP–Nd05) down to 3.43×10^{-42} cm^6/s (BAP–Nd50), with corresponding increases in the migration rate W_m, from 200 s^{-1} to 6310 s^{-1}, reflecting enhanced donor–donor migration at high concentrations. The critical distance R_{DA} for energy transfer was calculated using Eq. (6.11). Across the measured compositions, C_{DA} was consistently found to lie in the range of (3–8) $\times 10^{-42}$ cm^6/s. For comparison, the donor–donor energy migration microparameter C_{DD}, determined via the spectral overlap model, is approximately 3.8×10^{-39} cm^6/s (McCumber method) and 3.2×10^{-39} cm^6/s (Füchtbauer–Ladenburg method), with critical distances around 10.4 Å and 10.2 Å, respectively.

This disparity—nearly three orders of magnitude—highlights the predominant role of migration-assisted cross-relaxation in Nd^{3+} self-quenching at higher dopant concentrations in barium–alumino–metaphosphate glasses [32, 33].

References

1. R. Shi, A.-V. Mudring, ACS Materials Lett. **4**, 1882 (2022)
2. A. Kiel, in (1964), p. 765.

3. M.J. Weber, Phys. Rev. **171**, 283 (1968)
4. J.F.M. Dos Santos, V.S. Zanuto, A.C.C. Soares, E. Savi, L.A.O. Nunes, M.L. Baesso, T. Catunda, J. Lumin. **240**, 118430 (2021)
5. J. Day, S. Senthilarasu, T.K. Mallick, Sol. Energy Mater. Sol. Cells **199**, 83 (2019)
6. J. Drabik, R. Kowalski, L. Marciniak, Sci. Rep. **10**, 11190 (2020)
7. J.F.M. Dos Santos, N.G.C. Astrath, M.L. Baesso, L.A.O. Nunes, T. Catunda, J. Lumin. **202**, 363 (2018)
8. C.B. Layne, W.H. Lowdermilk, M.J. Weber, Phys. Rev. B **16**, 10 (1977)
9. B. Mitchell, H. Austin, D. Timmerman, V. Dierolf, Y. Fujiwara, Nanophotonics **10**, 851 (2020)
10. J.I. Eldridge, J. Lumin. **214**, 116535 (2019)
11. P. Zdeb, N. Rebrova, P.J. Dereń, J. Phys. Chem. Lett. **15**, 9356 (2024)
12. T. Kushida, J. Phys. Soc. Jpn. **34**, 1318 (1973)
13. L. Agazzi, K. Wörhoff, M. Pollnau, J. Phys. Chem. C **117**, 6759 (2013)
14. B. Liao, (ed.), Nanoscale Energy Transport: Emerging Phenomena, Methods and Applications (IOP Publishing, 2020).
15. L.A. Riseberg, M.J. Weber, in *Progress in Optics* (Elsevier, 1977), pp. 89–159.
16. J.M.F. Van Dijk, M.F.H. Schuurmans, J. Chem. Phys. **78**, 5317 (1983)
17. N. Boetti, D. Pugliese, E. Ceci-Ginistrelli, J. Lousteau, D. Janner, D. Milanese, Appl. Sci. **7**, 1295 (2017)
18. B. Faure, W. Blanc, B. Dussardier, G. Monnom, J. Non-Cryst. Solids **353**, 2767 (2007)
19. M. Kaczkan, M. Malinowski, Materials **14**, 3831 (2021)
20. D.L. Dexter, J. Chem. Phys. **21**, 836 (1953)
21. Th. Förster, Ann. Phys. **437**, 55 (1948)
22. B.W. Van Der Meer, in *FRET—Förster Resonance Energy Transfer*, edited by I. Medintz and N. Hildebrandt, 1st ed. (Wiley, 2013), pp. 23–62.
23. J.A. Caird, A.J. Ramponi, P.R. Staver, J. Opt. Soc. Am. B **8**, 1391 (1991)
24. H.C. Çamiçi, T. Guérineau, V.A.G. Rivera, R.F. Falci, S. LaRochelle, Y. Messaddeq, Sci. Rep. **13**, 8835 (2023)
25. O.L. Malta, Chem. Phys. Lett. **87**, 27 (1982)
26. A.I. Burshteĭn, Sov. Phys. Usp. **27**, 579 (1984)
27. M. Yokota, O. Tanimoto, J. Phys. Soc. Jpn. **22**, 779 (1967)
28. D. Wang, Y. Guo, Physica B **404**, 55 (2009)
29. P. Villanueva-Delgado, K.W. Krämer, R. Valiente, M. De Jong, A. Meijerink, Phys. Chem. Chem. Phys. **18**, 27396 (2016)
30. A.D. Sontakke, K. Biswas, A.K. Mandal, K. Annapurna, Appl. Phys. B **101**, 235 (2010)
31. C. Guo (ed.), *Handbook of Laser Technology and Applications*, 0 ed. (CRC Press, 2021).
32. A.S.S. De Camargo, C. Jacinto, T. Catunda, L.A.O. Nunes, Appl. Phys. B **83**, 565 (2006)
33. C. Jacinto, S.L. Oliveira, L.A.O. Nunes, J.D. Myers, M.J. Myers, T. Catunda, Phys. Rev. B **73**, 125107 (2006)

Chapter 7
Refractive Index

This chapter explores the spectral dependence of the refractive index in RE-doped glassy materials by combining classical electromagnetic models, fractional approaches, and experimental techniques. Dispersion-tuning models are discussed, as well as the structural and optical implications of the incorporation of REI^{3+} into glassy matrices.

The refractive index is a fundamental optical property that determines how light propagates through a material [1–3]. The refractive index of an optical material is determined by two main types of absorption: one due to electronic oscillations of resonance transitions in the UV region and the other due to the material's lattice vibration [4, 5]. In glasses doped with rare earth (RE), the refractive index is influenced not only by the composition of the host matrix but also by the type and concentration of dopant ions [6, 7]. These ions introduce localized polarization effects and modify the dielectric response, affecting both the real and imaginary parts of the refractive index [6, 8]. The spectral dependence of the refractive index can be described using classical electromagnetic oscillator models, the most common being the Drude–Lorentz model, which accounts for electronic resonances and damping mechanisms in the material [9, 10].

7.1 Drude–Lorentz Oscillator Model

The Drude–Lorentz model treats bound electrons in the material as classical harmonic oscillators subject to an external electric field $E(t)$ [11]. In this formalism, the ions are trapped around their equilibrium positions by a restoring force that is directly proportional to the displacement. The equation of motion is given by

$$\frac{d^2}{dt^2}x(t) + \gamma \frac{d}{dt}x(t) + \omega_0 x(t) = -\frac{eE(t)}{m} \tag{7.1}$$

G. Lozano C. et al., *A Brief Guide to Optical Spectroscopy of Trivalent Rare Earth Ions in Glasses*, Springer Series in Optical Sciences 247,
https://doi.org/10.1007/978-3-032-18501-3_7

where m and e are the effective mass and charge of the electron, γ is the damping constant, ω_o is the natural resonance frequency, and $x(t)$ is the displacement of the electron. Solving this equation in the frequency domain, assuming $E(t) = E_0 exp(i\omega t)$, yields the complex polarizability:

$$\chi(\omega) = \frac{Ne^2}{\varepsilon_0 m} \frac{1}{\omega^2{}_0 - \omega^2 - i\gamma\omega} \tag{7.2}$$

where N is the number of oscillators per unit volume ($N = N_A\rho/M$), ε_0 is the vacuum permittivity, and $\frac{Ne^2}{\varepsilon_0 m}$ is ω_p. The complex dielectric function is

$$\varepsilon(\omega) = \varepsilon_\infty + \chi(\omega) \tag{7.3}$$

with ε_∞ representing the contribution from higher-energy transitions not modeled explicitly. Furthermore, the complex refractive index is related to the dielectric function by

$$n(\omega) = \sqrt{\varepsilon(\omega)} = n'(\omega) + in''(\omega) \tag{7.4}$$

where $n'(\omega)$ is the real part, which governs the phase velocity of light, and $n'\prime(\omega)$ is the imaginary part, related to absorption losses. Separating the real and imaginary parts of $\varepsilon(\omega)$

$$\varepsilon'(\omega) = \varepsilon_\infty + \omega_p \frac{(\omega^2{}_0 - \omega^2)}{(\omega^2{}_0 - \omega^2)^2 + \gamma^2\omega^2} \tag{7.5}$$

$$\varepsilon''(\omega) = \varepsilon_\infty + \omega_p \frac{\gamma\omega}{(\omega^2{}_0 - \omega^2)^2 + \gamma^2\omega^2} \tag{7.6}$$

Figure 7.1a illustrates the frequency-dependent behavior of the dielectric permittivity and the resulting complex refractive index in a material modeled by the Drude–Lorentz oscillator formalism.

The response of the material is described in two complementary domains: the real part ε′ (green curve) indicates how strongly the material polarizes in response to an applied electric field, storing energy, and the imaginary part ε′′ (blue curve) reflects energy dissipation or absorption, peaking near the natural resonance of the materials. Figure 7.2 provides a physical interpretation by frequency region.

Then, the refractive index is computed from

$$n = \frac{1}{\sqrt{2}}\sqrt{\varepsilon' + \sqrt{\varepsilon'^2 + \varepsilon'\prime^2}} \tag{7.7}$$

$$k = \frac{1}{\sqrt{2}}\sqrt{\varepsilon' - \sqrt{\varepsilon'^2 + \varepsilon'\prime^2}} \tag{7.8}$$

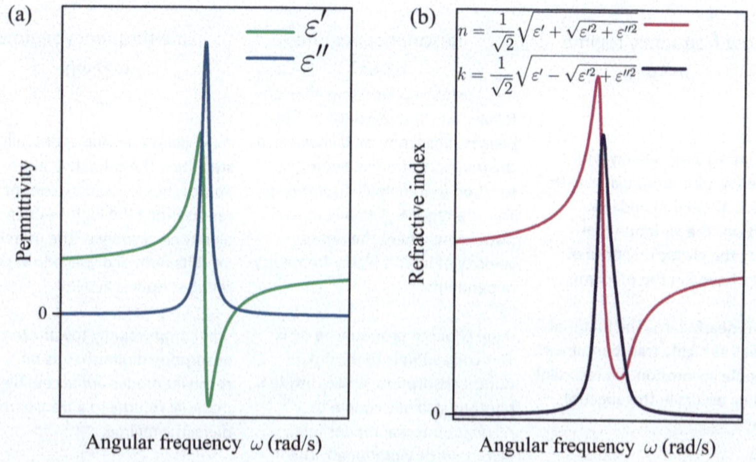

Fig. 7.1 Frequency-dependent behavior of the **a** complex dielectric permittivity and **b** refractive index in a resonant optical medium

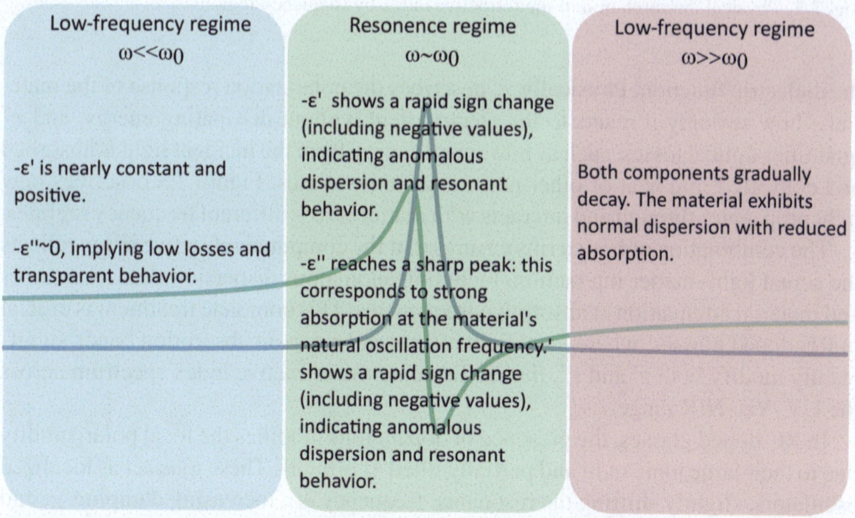

Fig. 7.2 Physical interpretation of the permittivity by frequency regions

We obtain the refractive index n (pink curve) and the extinction coefficient k (blue curve). $n(\omega)$ determines the phase velocity and refraction of light and $k(\omega)$ is proportional to the absorption coefficient $\alpha(\omega) = 4\pi k/\lambda$, representing attenuation of the electromagnetic wave.

Equations (7.7) and (7.8) define the magnitude of the complex refractive index, incorporating both the dispersive (real) and absorptive (imaginary) components of

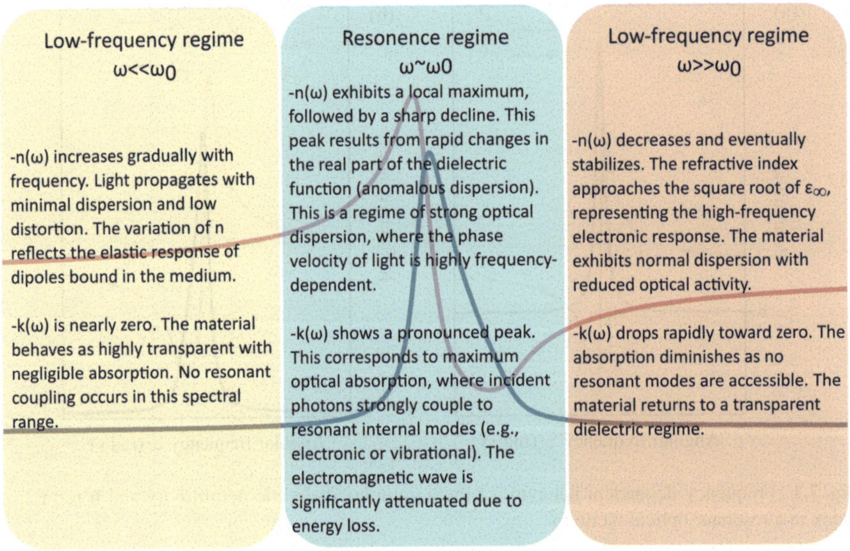

Fig. 7.3 Physical interpretation of the refractive index by frequency regions

the dielectric function. Physically, ε' describes the polarization response of the material—how strongly it reacts to the electric field without dissipating energy, and ε'' quantifies optical losses, such as how much energy from the incident light is absorbed and converted into heat or other non-radiative processes. Figure 7.3 describes how light propagates through and interacts with matter across different frequency regimes.

The combination of both terms ensures that the computed refractive index reflects the actual light–matter interaction, including anomalous dispersion near resonances and material attenuation at absorptive frequencies. This complete treatment is crucial in RE-doped glasses, where local field effects and resonant absorption bands significantly modify both ε' and ε'', thus influencing the refractive index spectrum across the UV–Vis–NIR range.

In RE-doped glasses, the presence of dopant ions modifies the local polarizability due to their large ionic radii and partially filled 4f orbitals. These ions act as localized oscillators, slightly shifting the resonance frequency ω_0, increasing damping γ, and enhancing the polarizability. This results in an increase in the refractive index with dopant concentration, broader absorption bands, especially in the near-infrared and enhanced nonlinear optical responses at resonant wavelengths.

7.1.1 Drude–Lorentz Model with Fractional Derivative

Further formalisms for the Drude–Lorentz model have been developed. The incorporation of a damping mechanism proportional to the ion velocity, as suggested by

Gonçalves [12] and El-Mallawany [13], allows the application of the generalized Drude–Lorentz model with fractional derivative operators, expressed as

$$\frac{d^2}{dt}x(t) + \gamma\frac{d^\alpha}{dt^\alpha}x(t) + \omega_0^2 x(t) = \frac{E_0 q}{m}e^{i\omega t} \tag{7.9}$$

where d^α/dx^α is the fractional derivative. Equation (7.8) could be treated as a special case of the Langevin equation, but, in the present case only, we use the special case of the fractional motion equation to investigate electro-optical properties [12]. According to the Caputo fractional derivative operator [14] is defined as

$$\frac{d^\alpha}{dt^\alpha}u(t) = \Gamma(1-\alpha)^{-1}\int_{t_o}^{t}(t-t_0)^{-\alpha}\frac{u(t')}{dt'}dt \tag{7.10}$$

where $0 < \alpha < 1$ and $\Gamma(1-\alpha)$ is the gamma function, and in addition the Fourier transform of Eq. (7.10) could be described as

$$F\left\{\frac{d^\alpha}{dt^\alpha}u(t)\right\} = (-i\omega)^\alpha U(\omega) \tag{7.11}$$

Applying Eqs. (7.2) and (7.3) to Eq. (7.10), we obtain the polarizability in the frequency space:

$$X(\omega,\alpha) = \frac{\pi mc^2}{e^2 N\phi}\left(\frac{1}{\omega_0^2+\omega^2+(-i)^\alpha\gamma\omega^\alpha}\right) \tag{7.12}$$

Here ϕ is the unitless oscillator strength, $\omega = 2\pi f$ is angular frequency, the product $N\phi$ is known as the dispersion and considering $\omega_p^2 = \frac{\pi mc^2}{e^2 N\phi}$. Considering that $(-i)^\alpha = cos\frac{\pi\alpha}{2} + isin\frac{\pi\alpha}{2}$, we obtain

$$X(\omega,\alpha) = \frac{\omega_p^2\left[\omega_0^2-\omega^2+\gamma\omega^\alpha cos\frac{\pi\alpha}{2}\right]}{\left[\omega_0^2+\omega^2+\gamma\omega^\alpha cos\frac{\pi\alpha}{2}\right]^2+\left[\gamma\omega^\alpha sin\frac{\pi\alpha}{2}\right]^2} - i\frac{\gamma\omega_p^2\omega^\alpha sin\frac{\pi\alpha}{2}}{\left[\omega_0^2+\omega^2+\gamma\omega^\alpha cos\frac{\pi\alpha}{2}\right]^2+\left[\gamma\omega^\alpha sin\frac{\pi\alpha}{2}\right]^2} \tag{7.13}$$

Taking into account the induced dipole moment (μ) and the polarization per unit volume:

$$X(\omega) = \mu(\omega)E(\omega) \tag{7.14}$$

$$P(\omega) = NX(\omega)E(\omega) = \varepsilon\chi(\omega)E(\omega) \tag{7.15}$$

We obtain the generalized permissivity:

$$\varepsilon(\omega, \alpha) = \varepsilon_\infty + X(\omega, \alpha) \tag{7.16}$$

And according to Reitz, Mildford, and Christy, the refractive index is described as

$$n^2 = \frac{1}{2}\left[Re(\varepsilon) + \sqrt{Re(\varepsilon)^2 + Im(\varepsilon)^2}\right] \tag{7.17}$$

where the real part $Re(\varepsilon)$ and imaginary part $Im(\varepsilon)$ are

$$Re(\varepsilon) = \varepsilon_\infty + \frac{\omega_p^2\left[\omega_0^2 - \omega^2 + \gamma\omega^\alpha cos\frac{\pi\alpha}{2}\right]}{\left[\omega_0^2 + \omega^2 + \gamma\omega^\alpha cos\frac{\pi\alpha}{2}\right]^2 + \left[\gamma\omega^\alpha sin\frac{\pi\alpha}{2}\right]^2} \tag{7.18}$$

$$Im(\varepsilon) = \frac{\gamma\omega_p^2\omega^\alpha sinsin\frac{\pi\alpha}{2}}{\left[\omega_0^2 + \omega^2 + \gamma\omega^\alpha cos\frac{\pi\alpha}{2}\right]^2 + \left[\gamma\omega^\alpha sin\frac{\pi\alpha}{2}\right]^2} \tag{7.19}$$

Considering that $\omega_j = 2\pi f_j$, where $j = 0, p.$. Equations (7.10) and (7.11) can be written as

$$Re(\varepsilon) = \varepsilon_0 + \frac{4\pi^2 f_p^4\left[(2\pi)^\alpha\gamma cos(\frac{\pi\alpha}{2})f^\alpha - 4\pi^2 f^2 + 4\pi^2 f_0^2\right]}{(2\pi)^{2\alpha}\gamma^2 sin(\frac{\pi\alpha}{2})f^{2\alpha} + \left[(2\pi)^\alpha\gamma cos(\frac{\pi\alpha}{2})f^\alpha - 4\pi^2 f^2 + 4\pi^2 f_0^2\right]^2} \tag{7.20}$$

$$Im(\varepsilon) = \frac{(2\pi)^{\alpha+2}\gamma sin(\frac{\pi\alpha}{2})f^\alpha f_p^2}{(2\pi)^{2\alpha}\gamma^2 sin(\frac{\pi\alpha}{2})f^{2\alpha} + \left[(2\pi)^\alpha\gamma cos(\frac{\pi\alpha}{2})f^\alpha - 4\pi^2 f^2 + 4\pi^2 f_0^2\right]^2} \tag{7.21}$$

Equation (7.13) delineates the spectral changes in the refractive index, which may depend on the α parameter of the fractional derivative. The α parameter can be linked to the degree of lattice irregularity, such as possible defects induced by the composition of glasses. Consequently, this parameter is connected to the maximum and minimum values of the refractive index, which may be governed by anomalous dispersion. γ represents the parameter reflecting the damping of the oscillator, which can be interpreted as dependent on lattice vibration.

7.1.2 Techniques for Refractive Index Determination

The refractive index is a fundamental optical parameter that governs light propagation, reflection, and confinement in dielectric media [15, 16]. Accurate measurement of this quantity is essential for the design of photonic devices, particularly in the context of RE-doped glasses, where the presence of dopant ions modifies the electronic structure and thus alters the optical response. Several experimental techniques are available for determining the refractive index of optical materials, each with its advantages and limitations depending on the geometry of the sample, the spectral range, and the precision requirements [17]. Common methods include:

- Spectroscopic methods (e.g., transmission via the Nussbaumer method) are useful for broad-spectrum dispersion fitting but are limited by model assumptions and sample thickness. Inaccurate calculation of refractive indices occurs when the transmission spectrum shows peaks of RE transitions [18].
- Ellipsometry measures change in polarization upon reflection; very accurate but complex and highly sensitive to surface roughness and model selection [19, 20].
- Abbe refractometry is fast and simple but is suitable only for transparent bulk materials at discrete wavelengths. It does not measure refractive indices greater than 2 [21].
- Prism coupling combines precision, non-destructive testing, and suitability for thin films or bulk waveguides, making it highly versatile [22].

Among the available techniques, the prism coupling method stands out as one of the most ideal methods for measuring the refractive index of REI-doped glasses. Its unique combination of high sensitivity, non-invasive contact, and spectral flexibility makes it particularly attractive for studying dispersion in optical materials with complex compositions.

7.1.3 Prism Coupling

The prism coupling method provides a direct measurement of the effective refractive index without requiring assumptions about absorption or film thickness [23]. The technique is compatible with multiple wavelengths, enabling analysis of chromatic dispersion and fitting of models such as Cauchy or Sellmeier. The method is non-destructive, making it appropriate for fragile or high-value RE-doped samples.

The instrumentation for this method can be seen in Fig. 7.4. It involves bringing the sample under analysis (with refractive index n) into contact with the base of a prism using a coupler. Then, a laser beam enters the prism (with refractive index n') at a variable angle of incidence θ and travels inside the prism until it reaches the sample surface, where it reflects entirely toward the photodetector. The system rotates around its axis until the angle of incidence is less than the critical angle θ_c, which is detected by the photodetector and interpreted as a sharp drop in the intensity curve. The advantages of this technique are that it is non-destructive and applicable to thin films, waveguides and even bulk glasses [24]. It is also compatible with a wide spectral range, depending on the laser sources (e.g., 532 nm, 632.8 nm, 1310 nm, and 1550 nm).

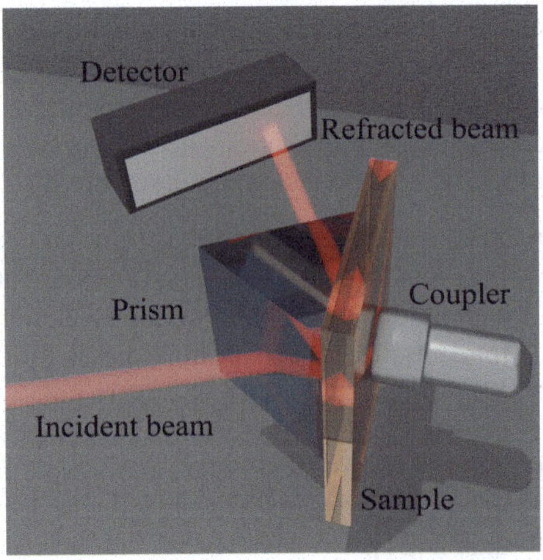

Fig. 7.4 Schematic representation of the prism coupling method

7.1.4 *Fitting Models*

Modeling the refractive index as a function of wavelength or photon energy is fundamental to understanding the optical behavior of dielectric materials. In rare-earth-doped glasses, where electronic structure, local polarizability, and structural disorder all contribute to dispersion, selecting an appropriate fitting model is essential for extracting physical parameters and predicting performance in unmeasured spectral regions. This section outlines the most commonly used models for fitting the refractive index, their theoretical foundations, applicability ranges, and limitations.

7.1.5 *Sellmeier Model*

In that line of thought, the Sellmeier equation is an empirical relationship between the wavelength λ and the refractive index n of a transparent medium, used to determine the dispersion of light in an optically transparent material. It was deduced in 1871 by W. Sellmeier from deductions obtained in A. Cauchy's works to model media with dispersion [25]. The single-term Sellmeier equation is

$$n^2(\lambda) = 1 + \frac{A\lambda^2}{\lambda^2 - B} \tag{7.22}$$

Here A and B are the Sellmeier coefficients determined from the experimental data. These coefficients are available in databases for many optical materials, such as in the case of tellurite glasses reported by Ghosh, 1997 [26]. This expression accounts for electronic resonances beyond the visible or UV range and is derived from the classical Lorentz oscillator model under non-absorbing conditions. The multi-term Sellmeier equation is

$$n^2(\lambda) = 1 + \sum_i \frac{A_i\lambda^2}{\lambda^2 - B_i} \tag{7.23}$$

To study the influence of Rei doping on refractive index dispersion, we analyzed the glassy system (70–x)% TeO_2–20% ZnO–10% K_2O–x% Yb_2O_3, where x = 0.0, 0.3, 0.6, 0.9, 1.2, and 1.5 mol% [27]. For all compositions, as the Yb_2O_3 concentration increases, the refractive index systematically decreases across the entire spectral range, see Fig. 7.5. The inset highlights the linear decrease of the refractive index at 972 nm as a function of Yb_2O_3 molar content. This trend is of particular significance for photonic applications operating near 1 μm, such as Yb^{3+}-based lasers and amplifiers, where precise knowledge of the refractive index is essential for mode confinement and design of optical waveguides [28, 29].

The addition of Yb_2O_3 into the glass network partially replaces structural TeO_2 units. Yb^{3+} ions, having lower polarizability than Te^{4+}, contribute to an overall decrease in mean electronic polarizability, thus reducing the refractive index [30–32]. Although Yb_2O_3 may slightly increase the glass density, the refractive index tends to decrease with Yb^{3+} content, leading to lower refractive indices (Fig. 7.5). RE doping

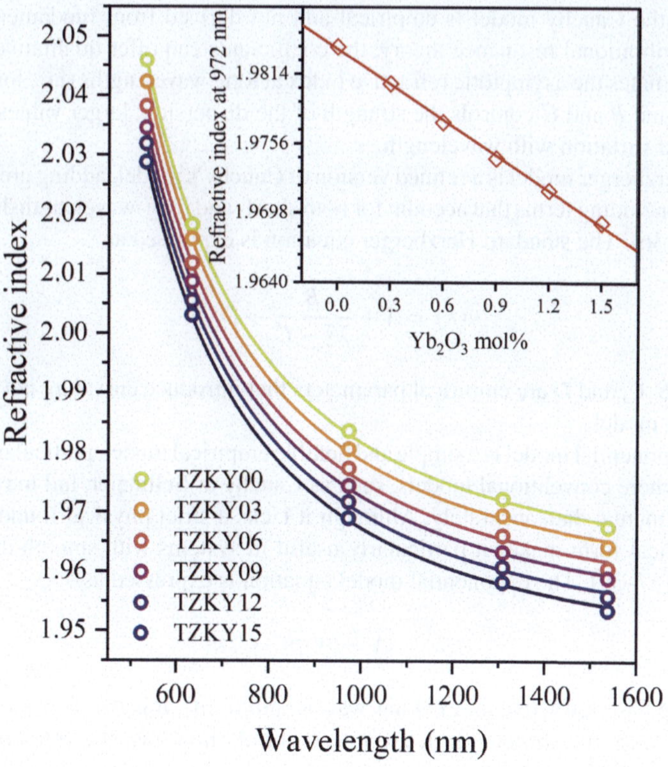

Fig. 7.5 Dispersion of the refractive index for (70–x)% TeO_2–20% ZnO–10% K_2O–x% Yb_2O_3 glasses fitted using the Sellmeier model

often induces network depolymerization (formation of non-bridging oxygens), which can alter local bond polarizability. In this system, the effect appears to be a net reduction in the refractive index, possibly due to an increase in structural symmetry or reduced distortion. The ability to adjust the refractive index linearly with Yb^{3+} concentration allows precise control over waveguide core–cladding index contrast, critical in the fabrication of integrated photonic devices [33, 34].

7.1.6 Other Models

The Cauchy model is one of the simplest empirical formulas for transparent materials in spectral regions without strong absorption [35]. The general form of the Cauchy equation is given by

$$(\lambda) = A + \frac{B}{\lambda^2} + \frac{C}{\lambda^4} + \cdots \tag{7.24}$$

where A, B, and C are empirical coefficients obtained by fitting experimental data. Although the Cauchy model is empirical and not derived from fundamental electronic or vibrational resonance theory, the coefficients can offer qualitative insight: A approximates the asymptotic refractive index at long wavelengths (i.e., low photon energies) and B and C controls the strength of the dispersion; larger values indicate more rapid variation with wavelength.

The Herzberger model is a refined version of Cauchy's model, adding greater flexibility by including terms that account for both short- and long-wavelength dispersion behavior [36]. The standard Herzberger equation is expressed as

$$n(\lambda) = A + \frac{B}{\lambda^2 - C} + D\lambda^2 \tag{7.25}$$

where A, B, C, and D are empirical parameters that introduce curvature not captured in simpler models.

The exponential model is a simple and flexible empirical model, particularly useful in cases where conventional models, such as Cauchy or Sellmeier, fail to reproduce the experimental data accurately. Although it lacks a strict physical foundation, its mathematical form makes it particularly useful in systems with smooth dispersion behavior [37–39]. The exponential model equation is expressed as

$$n(\lambda) = A + Bexp(-C/\lambda) \tag{7.26}$$

A is the base refractive index (long-wavelength limit), B is the dispersion amplitude, and C is the dispersion decay rate (controls how fast the refractive index decreases with λ).

The Wemple–DiDomenico model is a semi-empirical dispersion model that relates the refractive index of a dielectric material to its electronic band structure through a single effective oscillator [40]. Unlike purely empirical models such as Cauchy or exponential equations, this model provides a physically meaningful description of how $n(\lambda)$ it changes with photon energy, especially in the transparent region below the absorption edge. The refractive index n as a function of photon energy $E = h\nu$ (in eV) is given by

$$n^2(E) = 1 + \frac{E_d E_0}{E^2{}_0 - E^2} \tag{7.26}$$

E is the photon energy in eV, $E = 1240/\lambda$, E_0 is the single oscillator energy in eV that is correlates with the optical bandgap, and E_d is the dispersion energy (in eV), related to electronic structure of the material that is proportional to the product of the coordination number, valence, the density of the polarizable bonds.

Figure 7.6 presents refractive index dispersion curves for three distinct rare-earth-doped glass systems: (a) TGZN_Er (tellurite-germanate doped with 1% mol Er_2O_3), and (b) GBNL_Tm_Er (borogermanate doped with 0.5% mol Er_2O_3 and 0.5% mol Tm_2O_3). The experimental data points (purple symbols) are fitted using five different dispersion models: Sellmeier (yellow), Cauchy (green), Herzberger (cyan), Exponential (blue), and Wemple–DiDomenico (dark blue). For all systems, the refractive index $n(\lambda)$ shows a monotonic decrease with increasing wavelength, characteristic of normal dispersion in glasses. The fits for each model closely follow the experimental data, but subtle differences in accuracy and fitting behavior emerge, especially for TGZN_Er and GBNL_Tm_Er compositions.

TGZN_Er glass exhibits the lowest dispersion curvature among the samples analyzed, especially at short wavelengths (400–600 nm), where the dispersions between models become more pronounced [41]. The Cauchy and Herzberger models provide a better fit in the visible region; the Herzberger model achieves a better curvature fit due to its greater flexibility. Although the exponential model accurately follows the general trend, it slightly softens the curvature near 500–600 nm; the Wemple–DiDomenico model performs well throughout the range. The composition of the TGZN host exhibits greater sensitivity to structural changes induced by the incorporation of rare earth, so more flexible models such as Herzberger or Wemple–DiDomenico are required to capture subtle variations.

In the borogermanate glass GBNL_Tm_Er, the scattering profile is comparatively flatter. The Sellmeier model provides a globally consistent fit but slightly underestimates the experimental values at the shortest wavelengths. Both the Cauchy and Herzberger models closely follow the experimental points; Herzberger shows a slight advantage at longer wavelengths because of its quadratic term. The exponential model fits well, especially in this system, where the scattering slope is relatively smooth. The Wemple–DiDomenico fit remains robust, retaining physical significance and capturing the subtle slope changes introduced by the co-doping of Tm^{3+} and Er^{3+} because its parameters are related to the electronic structure of the material. The

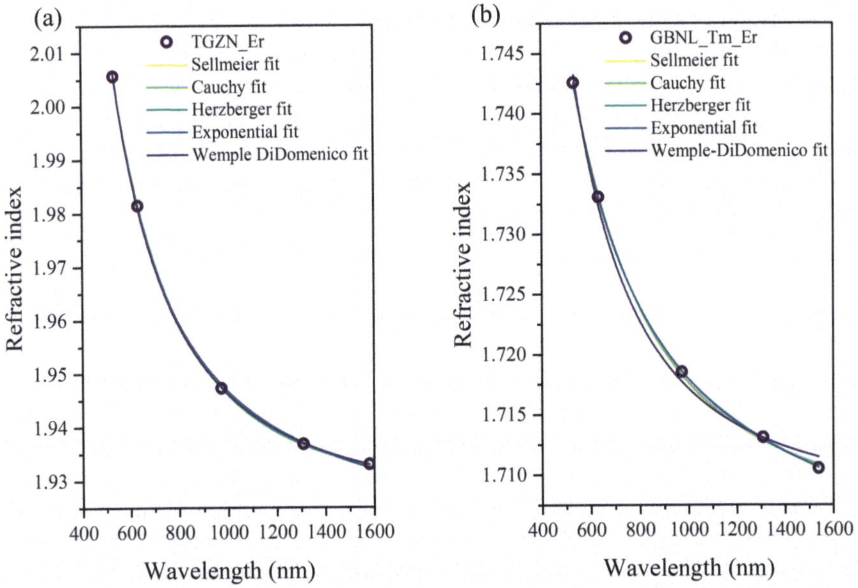

Fig. 7.6 Refractive index dispersion of **a** TGZN_Er and **b** GBNL_Tm_Er glasses, fitted with Sellmeier, Cauchy, Herzberger, Exponential, and Wemple–DiDomenico models

observed variations in the fits to GBNL_Tm_Er glass reflect the combined effects of two rare-earth ions with distinct electron configurations, compared to those observed in TGZN_Er glass.

In the context of refractive index fitting models, the appearance of negative values in the fitting coefficients can have different origins and meanings, depending on the model and the physical context.

The conceptual diagram, Fig. 7.7, summarizes the causal relationship between RE doping and the modification of the refractive index $n(\lambda)$ in glass matrices. Each block in the sequence reflects a distinct physical mechanism that, when combined, explains the optical behavior of RE-doped systems.

Here, we presented a comprehensive study of the refractive index in RE-doped glasses, combining both theoretical modeling and experimental considerations. In these systems, the refractive index is strongly influenced not only by the host glass matrix but also by the type and concentration of dopant ions, which modify local polarizability, introduce resonance effects, and alter the dielectric function of the material. We began by discussing the classical Drude–Lorentz oscillator model, which provides a physical foundation for understanding the frequency-dependent dielectric response of bound electrons. Extensions of this model using fractional calculus were also introduced to account for complex damping mechanisms and structural disorder, which are especially relevant in amorphous and glass systems.

Experimental methods for determining the refractive index were reviewed in detail, highlighting their advantages and limitations. Among them, the prism coupling

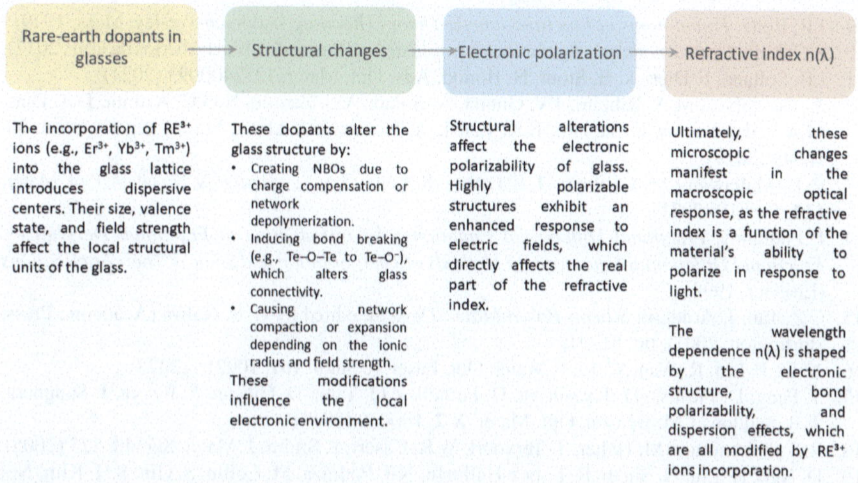

Fig. 7.7 Conceptual diagram illustrates the causal relationship between rare-earth ion doping and changes in the refractive index $n(\lambda)$ in glass materials

technique was identified as particularly suited for RE-doped glasses because of its precision, non-destructive nature, and compatibility with a broad spectral range. A series of dispersion fitting models—including Sellmeier, Cauchy, Herzberger, Exponential, and Wemple–DiDomenico—was examined to model the spectral dependence of the refractive index. The theoretical basis, fitting flexibility, and physical interpretability of each model were analyzed.

Finally, the ability to tune the refractive index through REI doping was emphasized as a valuable strategy in the design of photonic devices. The linear variation of the refractive index with dopant content enables precise control over optical contrast, critical for waveguides and amplifiers. The choice of dispersion model should, therefore, balance fitting accuracy, physical significance, and intended application.

References

1. T.S. Soliman, Ceram. Int. **51**, 25408 (2025)
2. P. Loiko, N. Ismail, J.D.B. Bradley, M. Götelid, M. Pollnau, J. Non-Cryst, Solids **476**, 95 (2017)
3. J. Wei, J.M. Murray, J.O. Barnes, D.M. Krein, P.G. Schunemann, S. Guha, Opt. Mater. Express **8**, 485 (2018)
4. R. El-Mallawany (ed.), *Tellurite Glass Smart Materials* (Springer International Publishing, Cham, 2018)
5. R.K. Verma, in *Ceram. Sci. Eng.*, edited by K.P. Misra, R.D.K. Misra (Elsevier, 2022), pp. 49–62.
6. S. Yonezawa, S. Nishibu, M. Leblanc, M. Takashima, J. Fluor. Chem. **128**, 438 (2007)
7. L.R.P. Kassab, M.J.V. Bell, in *Lanthan.-Based Multifunct. Mater.*, edited by P. Martín-Ramos and M. Ramos Silva (Elsevier, 2018), pp. 263–289.
8. A. Ratep, I. Kashif, Radiat. Phys. Chem. **237**, 113095 (2025)

9. J.R. Reitz, *Foundations of Electromagnetic Theory (Reading* (Addison-Wesley, Mass, 1979)
10. S.C. Zilio, *Óptica moderna: fundamentos e aplicações* (Portal de Livros Abertos da USP, 2017).
11. I.B. Soltane, F. Dierick, B. Stout, N. Bonod, Adv. Opt. Mater. **12**, 2400093 (2024)
12. A. Gonçalves, M.A. Ribeiro, J.V. Gunha, A. Somer, V.S. Zanuto, N.G.C. Astrath, D.T. Dias, M.A.F. dos Santos, C. Jacinto, E.K. Lenzi, A. Novatski, J. Mater. Sci. Mater. Electron. **30**, 16949 (2019)
13. R. El-Mallawany, M.A. Ribeiro, L.S. Lara, E.K. Lenzi, I.A.A. Alsadig, A. Novatski, Opt. Mater. **112**, 110810 (2021)
14. I. Podlubny, *Fractional Differential Equations: An Introduction to Fractional Derivatives, Fractional Differential Equations, to Methods of Their Solution and Some of Their Applications* (Elsevier, 1998).
15. C.Z. Tan, J. Arndt, in *Silicon-Based Mater. Devices*, edited by H. S. Nalwa (Academic Press, Burlington, 2001), pp. 51–91.
16. X. Li, F. Liu, R. Wan, W. Li, P. Wang, Opt. Laser Technol. **161**, 109216 (2023)
17. Y. Fang, D. Furniss, D. Jayasuriya, H. Parnell, Z.Q. Tang, D. Gibson, S. Bayya, J. Sanghera, A.B. Seddon, T.M. Benson, Opt. Mater. X **2**, 100030 (2019)
18. R.J. Nussbaumer, M. Halter, T. Tervoort, W.R. Caseri, P. Smith, J. Mater. Sci. **40**, 575 (2005)
19. D. Ngo, H. Liu, N. Sheth, R. Lopez-Hallman, N.J. Podraza, M. Collin, S. Gin, S.H. Kim, Npj Mater. Degrad. **2**, 20 (2018)
20. I. Fabijanić, P. Pervan, B. Okorn, J. Sancho-Parramon, V. Janicki, Appl. Opt. **59**, A69 (2020)
21. A. García-Valenzuela, H. Contreras-Tello, Opt. Lett. **38**, 775 (2013)
22. Opt. Commun. **492**, 126984 (2021)
23. R. Ulrich, R. Torge, Appl. Opt. **12**, 2901 (1973)
24. V.A.G. Rivera, E.F. Chillcce, E. Rodriguez, C.L. Cesar, L.C. Barbosa, J. Non-Cryst, Solids **352**, 363 (2006)
25. S. Heshmati, H. Taleb, A. Rahmani, Optik **172**, 851 (2018)
26. G. Ghosh, Appl. Opt. **36**, 1540 (1997)
27. G. Lozano C., J.F.M. dos Santos, L.G. Merízio, M. de Oliveira Jr., Y. Messaddeq, E. Marega Jr., V.A.G. Rivera, Mater. Res. Bull. **189**, 113458 (2025)
28. G. Venkataiah, P. Babu, I.R. Martín, K. Venkata Krishnaiah, K. Suresh, V. Lavín, C.K. Jayasankar, J. Non-Cryst. Solids **479**, 9 (2018)
29. M.A. Merzliakov, V.V. Kouhar, G.E. Malashkevich, E.V. Pestryakov, Opt. Mater. **75**, 142 (2018)
30. L. Zhang, Y. Xia, X. Shen, R. Yang, W. Wei, Opt. Mater. **75**, 1 (2018)
31. J. Chacaliaza-Ricaldi, V.A.G. Rivera, I.C. Pinto, Y. Messaddeq, E. Marega, J. Non-Cryst, Solids **638**, 123063 (2024)
32. L.R.P. Kassab, R.A. Kobayashi, M.J.V. Bell, A.P. Carmo, T. Catunda, J. Phys. Appl. Phys. **40**, 4073 (2007)
33. A.S. Pinheiro, A.M. Freitas, G.H. Silva, M.J.V. Bell, V. Anjos, A.P. Carmo, N.O. Dantas, Chem. Phys. Lett. **592**, 164 (2014)
34. L. Zhang, T. Xue, D. He, M. Guzik, G. Boulon, Opt. Express **23**, 1505 (2015)
35. E. Stoumbou, I. Stavrakas, G. Hloupis, A. Alexandridis, D. Triantis, K. Moutzouris, Opt. Quantum Electron. **45**, 837 (2013)
36. M. Herzberger, JOSA **40**, 424 (1950)
37. J.M. Nichols, J. Opt. **22**, 045601 (2020)
38. J. Lumeau, L.B. Glebov, Opt. Mater. Express **3**, 95 (2013)
39. S.E. Savotchenko, Waves Random Complex Media **0**, 1 (n.d.)
40. M. DiDomenico, Appl. Opt. **11**, 652 (1972)
41. J. Chacaliaza-Ricaldi, G.L.C., J.L.C.H., F.A. Ferri, A.D. Rodrigues, Y. Messaddeq, V.A.G. Rivera, E. Marega, J. Non-Cryst. Solids **616**, 122471 (2023)

Chapter 8
Nonlinear Properties

Nonlinear optical (NLO) characterization techniques have emerged as essential tools to investigate the optical behavior of advanced materials, especially those doped with rare-earth ions (REIs) [1–4]. These materials exhibit a wide range of photonic applications, including laser systems, optical amplifiers, up-conversion devices, and optical limiters. Although linear optical measurements provide valuable information under low excitation conditions—such as absorption spectra, emission profiles, and refractive index—nonlinear techniques enable deeper exploration of intensity-dependent optical phenomena.

Among the various NLO methods, the Z-scan technique stands out as a powerful and widely used tool to investigate third-order nonlinearities in solid-state materials, glasses, crystals, and nanostructures [1, 2, 4, 5]. Introduced by Sheik-Bahae and colleagues in 1990, the Z-scan technique allows the simultaneous determination of the nonlinear refractive index (n_2) and the nonlinear absorption coefficient, using a single focused laser beam and a relatively simple experimental setup [4, 6].

In rare-earth-doped systems, the Z-scan technique is particularly useful due to the distinctive electronic configuration of REI [4, 7–9]. Their partially filled 4f orbitals are shielded from the crystal field by outer 5 s and 5p electrons, leading to sharp, well-defined energy levels that enable a variety of optical transitions. Under high-intensity excitation, these transitions can give rise to nonlinear responses through mechanisms such as multiphoton absorption, excited-state absorption, energy transfer up-conversion, and changes in population distributions among energy levels.

In this chapter, we focus specifically on the origin of the nonlinear refractive index associated with changes in the population of excited states induced by REI doping—a mechanism often referred to as the population lens effect. This phenomenon arises when intense laser irradiation modifies the distribution of ions between the ground and excited states, leading to a spatially varying refractive index. As a result, the material behaves like a self-induced lens, whose focusing or defocusing action depends on the population dynamics of the dopant ions [4, 10, 11].

G. Lozano C. et al., *A Brief Guide to Optical Spectroscopy of Trivalent Rare Earth Ions in Glasses*, Springer Series in Optical Sciences 247,
https://doi.org/10.1007/978-3-032-18501-3_8

This chapter introduces the fundamental concepts of nonlinear optics relevant to rare-earth-doped materials and presents the theoretical and experimental foundations of the Z-scan technique, with an emphasis on the population lens effect. We discuss both open-aperture (OA) and closed-aperture (CA) Z-scan configurations and explore how this technique can be applied to extract third-order susceptibility parameters [1, 3, 5]. Examples from recent studies will be presented to demonstrate the utility of the Z-scan technique in understanding nonlinear refractive behavior driven by excited-state population effects, thereby supporting the design of next-generation photonic materials and devices [12–14].

8.1 Theoretical Background of the Z-Scan Technique

The Z-scan technique [3, 5, 8] is a single-beam, third-order nonlinear optical method used to quantify both the nonlinear absorption and the nonlinear refractive index of a material. It is based on the principle of measuring the transmittance of a focused laser beam as the sample is translated along the propagation axis (z-axis) through the beam waist. As presented in Fig. 8.1, the method relies on intensity variations experienced by the sample at different positions relative to the focus, thus enabling the extraction of nonlinear optical coefficients from the transmittance profile.

In a typical Z-scan experiment, a laser beam—usually with a Gaussian spatial profile—is focused by a lens, and the sample is moved along the beam propagation direction (z-axis). The intensity of the beam varies with position z, reaching its maximum at the focal point. The intensity at any point along the z-axis can be expressed as [15, 16]

$$I(z) = \frac{2P_0}{\pi\omega(z)^2} = I_0\left(\frac{1}{1 + (z/z_0)^2}\right) \tag{8.1}$$

where P_0 is the laser power, $\omega(z)$ is the beam radius at position z, $z_0 = \pi w_0^2/\lambda$ is the Rayleigh range, and I_0 ($I_0 = 2P/\pi w_0^2$), is the peak intensity at focus.

To measure the nonlinear refractive index n_2, a partially closed aperture is placed in front of the detector, making the transmittance sensitive to beam distortion caused by intensity-dependent refractive index changes in the sample. The expression relating the normalized transmittance in the center of the beam with the z-displacement was obtained by Sheik-Bahae et al. [3]

$$T(x, \Delta\Phi_o) = 1 + \frac{4.x.\Delta\Phi_o}{(x^2 + 9)(x^2 + 1)} \tag{8.2}$$

where $x = z/z_0$. As the sample moves through the focus, the beam experiences self-focusing or self-defocusing effects, depending on the sign of n_2. These effects lead to a characteristic "peak-valley" or "valley-peak" transmittance pattern in the closed-aperture Z-scan trace.

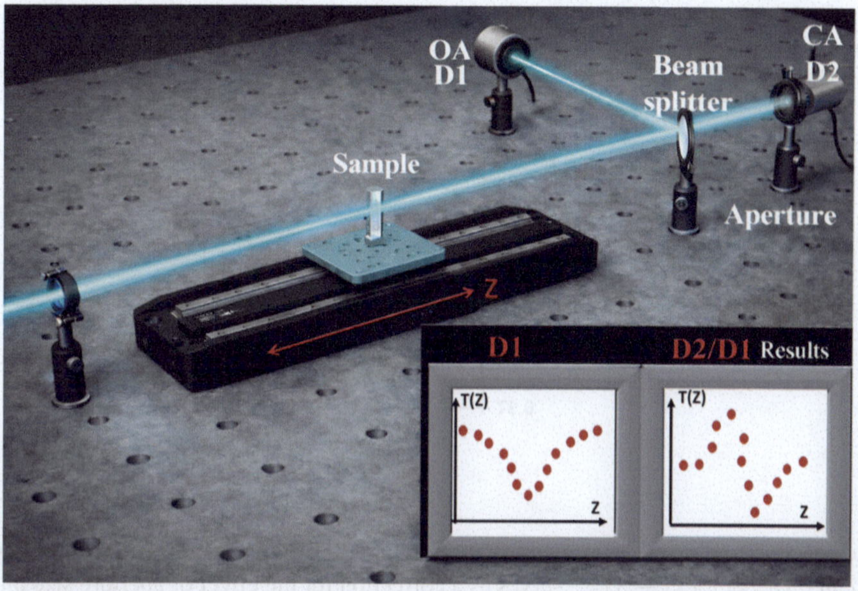

Fig. 8.1 Experimental setup of the Z-scan technique used to characterize the nonlinear optical properties of a sample. A laser beam is focused by a lens and directed onto the sample, which is translated along the z-axis relative to the focal plane. After the sample, the beam is split by a beam splitter. In the open-aperture (OA) configuration, detector D1 collects the entire transmitted beam, providing information primarily on nonlinear absorption. In the closed-aperture (CA) configuration, detector D2, positioned after an aperture, measures a portion of the transmitted beam, allowing the detection of nonlinear refraction effects. The inset graphs show typical transmittance T(z) curves as a function of the sample position z, with the ratio D2/D1 used to extract the refractive contribution

The phase shift induced by nonlinear refraction is given by

$$\Delta\Phi_0 = \frac{2\pi}{\lambda} L_{eff}\, n_2' I_0 \tag{8.3}$$

where L_{eff} represents the effective length of the sample, given by

$$L_{eff} = \frac{1-\exp(\alpha L)}{\alpha} \tag{8.4}$$

with α is the linear absorption coefficient, according to the Beer–Lambert law. For small phase shifts in a closed-aperture Z-scan measurement, the peak-to-valley transmittance difference, ΔT_{pv}, is proportional to the on-axis phase shift at the focus, $\Delta\Phi_o$, and can be estimated using the following expression [3, 11]:

$$\Delta T_{pv} = 0,406(1 - S)^{0,25}|\Delta\Phi_o|, \text{ with } |\Delta\Phi_o| \leq \pi \tag{8.5}$$

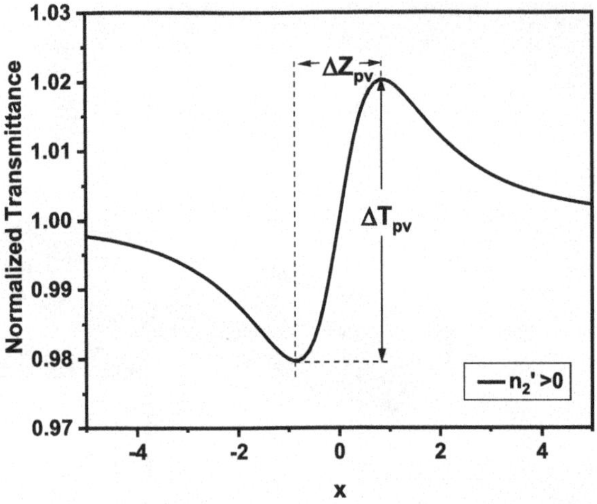

Fig. 8.2 Simulation of a typical closed-aperture Z-scan curve for $n_2' > 0$, assuming $\Delta\Phi_o$=0.1

where S is the linear transmittance of the aperture (i.e., the aperture fraction in the far field). The factor $(1 - S)^{0,25}$ serves as a correction for the effect of aperture size on the measured signal. The spatial separation between the peak and valley positions, ΔT_{pv}, can also be approximated by differentiating the transmittance function, leading to

$$\Delta z_{pv} \approx 1,7z_0 \tag{8.6}$$

In Fig. 8.2, we simulate the transmittance behavior based on Eq. (8.2), considering an on-axis phase shift $\Delta\Phi_o$=0.1. The parameters ΔT_{pv} and Δz_{pv} described in Eqs. (8.5) and (8.6), respectively, are highlighted.

On the other hand, performing a Z-scan with a fully open aperture (S = 1), it is possible to determine the nonlinear absorption coefficient [17], which is related to the imaginary part of the nonlinear refractive indexn_2''.. In this configuration, the normalized transmittance curve is symmetric with respect to the focal plane. As the sample passes through the region of highest intensity, nonlinear absorption increases, and the transmitted intensity reaches a minimum at the focus. This behavior can result from multiphoton absorption processes, such as two-photon absorption (TPA), three-photon absorption, or reverse saturable absorption.

The normalized transmittance as a function of the reduced position x = z/z$_0$ is given by [18]

$$T\left(x, \Delta\Phi_0''\right) = \frac{1}{2}\left(\frac{2\pi}{\lambda}\right)n_2''L_{eff}\frac{I_0}{1+x^2} = \frac{1}{2}\Delta\Phi_0''\frac{1}{1+x^2} \tag{8.7}$$

where $\Delta\Phi_0''$ represents the nonlinear phase shift associated with nonlinear absorption in the open-aperture configuration. A typical OA Z-scan curve is presented in Fig. 8.3.

Fig. 8.3 Simulation of typical open-aperture (OA) Z-scan curve

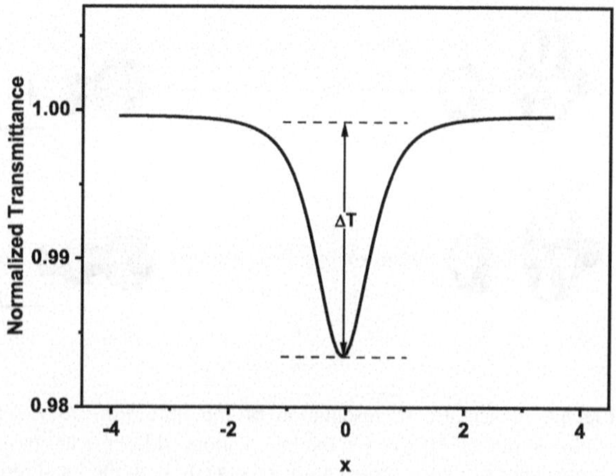

8.2 Time-Resolved Z-Scan for Slow Nonlinearities

In systems where nonlinear refractive index changes arise from slow processes—such as excited-state population dynamics in rare-earth-doped materials, the temporal response of the material becomes a crucial factor in the interpretation of Z-scan data. These slow nonlinearities, with characteristic response times (τ) on the order of hundreds of microseconds or more, typically originate from the gradual build-up of excited-state populations under excitation. In such cases, the induced refractive index change (Δn) evolves approximately as

$$\Delta n \approx N_{ex}(t) \approx N_t\left(1 - e^{-\frac{t}{\tau}}\right), \text{ with }, \tau = 100\,\mu s, \tag{8.8}$$

where $N_{ex}(t)$ is the population of the excited state at time t, N_t is the total dopant ion concentration, and τ is the characteristic lifetime or response time of the nonlinear effect. Given the relatively slow evolution of Δn, it is essential to resolve the signal temporally to distinguish these effects from instantaneous or static linear contributions.

A practical and widely adopted approach to isolating slow nonlinear contributions is to work with a normalized transmittance function, T(z), defined as the ratio between two measured signals at each z-position T(z) = Signal(t_f)/Signal(t_i), i.e., the transmitted intensity acquired immediately after the beam is unblocked and the transmitted intensity acquired collected just before the chopper blocks the beam again. This method effectively compares the initial (low-excitation) and final (steady-state) transmitted signals for each position along the z-axis.

This normalization offers a crucial advantage: it eliminates possible time-independent linear artifacts, such as those arising from poor surface polishing or slight deviations from parallelism in the sample. These effects influence both Signal(t_i) and

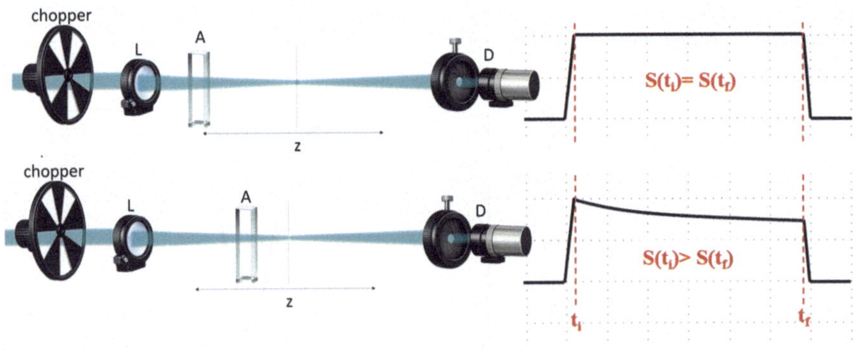

Fig. 8.4 Schematic representation of the time-resolved Z-scan technique for detecting slow nonlinear optical effects. On the left, a chopped laser beam passes through the sample (A), which is translated along the propagation axis (z) through the focal region (dashed line). The transmitted signal is collected by a detector (D) after passing through an aperture. The upper and lower panels correspond to two different sample positions: far from the focus (top) and near the focus (bottom). On the right are the corresponding temporal profiles of the detected signal. At positions far from the focus, the signal remains constant during the chopper opening time, giving $S(t_i) = S(t_f)$ and $T(z) = 1$. Near the focus, nonlinear effects become evident as the transmitted signal decreases over time, resulting in $S(t_i) > S(t_f)$ and $T(z) < 1$. This behavior reflects the accumulation of slow refractive nonlinearities, such as the population lens effect, over the duration of the laser pulse

Signal(t_f) equally and thus cancel out in the ratio T(z). In contrast, slow nonlinearities only develop over time and manifest distinctly in Signal(t_f), making them clearly detectable in the normalized transmittance, as illustrated schematically in Fig. 8.4.

The time-resolved Z-scan approach enables the detection of extremely small phase distortions—on the order of $\lambda/10^4$—and is therefore well suited to characterize weak but temporally cumulative nonlinear effects, such as those associated with population lensing. The minimum acquisition times for t_i and t_f are determined by the response times of both the optical chopper and the photodetectors (labeled D_1 and D_2 in Fig. 8.1), which must be sufficiently fast to resolve the temporal evolution of the nonlinear response.

The Z-scan technique offers several advantages: it is highly sensitive, requires only modest laser powers, and can be implemented with relatively simple optics. Moreover, the simultaneous measurement of both nonlinear refractive and absorptive properties from the same experimental data makes Z-scan particularly attractive for characterizing new photonic materials.

In rare-earth-doped systems, where excited-state population dynamics play a significant role [3, 11, 19], Z-scan measurements can reveal complex intensity-dependent behaviors associated with ESA and energy transfer processes. When combined with complementary time-resolved and spectroscopic techniques, Z-scan provides a comprehensive picture of the nonlinear optical response and its physical origin.

8.3 Examples of Rare-Earth Ions in Z-Scan Studies

In rare-earth-doped materials, Z-scan has been applied extensively to quantify third-order nonlinearities, highlighting the roles of specific REI and host matrices in phenomena like population lensing. Below are detailed illustrative cases:

$NaYF_4$:Yb, Er Core–Shell Nanocrystals Under CW Excitation at 532 nm

$NaYF_4$:Yb, Er@$NaYF_4$ up-converting nanocrystals probed by Z-scan under continuous-wave 532 nm excitation show a decrease in both β ($\beta \propto N_t \Delta\sigma / I_s$, where $\Delta\sigma$ population difference between ground and excited states and, I_s is the pump saturation intensity) and n_2 with increasing excitation power. At 30 mW, $\beta \approx 6.1 \times 10^{-4}$ cm/W and $n_2 \approx 8.9 \times 10^{-8}$ cm^2/W; at 120 mW, β drops to $\approx 1.1 \times 10^{-4}$ cm/W, while n_2 increases to $\approx 10.6 \times 10^{-8}$ cm^2/W [20]. These results emphasize the influence of excited-state absorption, saturation, and thermal effects—all of which are consistent with population-dependent nonlinearity.

Nd^{3+}-Doped Glasses and Ruby Crystals Investigated by Time-Resolved CW Z-scan

In a comparative study employing continuous-wave (cw) time-resolved Z-scan measurements, different nonlinear refractive mechanisms were identified in Nd^{3+}-doped glass samples and ruby (Al_2O_3:Cr^{3+}) crystals [4]. The approach, based on the acquisition of normalized transmittance curves, enabled discrimination between fast (electronic) and slow (thermal or population-based) nonlinear contributions by varying the acquisition time during the chopper cycle.

For Nd^{3+}-doped glasses, a clear temporal evolution of the transmittance signal was observed, indicative of a slow nonlinear refractive index variation. This behavior was attributed to the gradual build-up of excited-state populations (population lens effect), which modifies the local refractive index over microsecond timescales. In contrast, ruby crystals exhibited an almost instantaneous response, dominated by electronic Kerr nonlinearity, with no significant temporal evolution in the normalized transmittance.

These results highlight the importance of considering temporal resolution in Z-scan experiments involving rare-earth-doped materials, especially when slow mechanisms—such as excited-state population dynamics or thermal lensing—are expected to contribute to the observed nonlinearity. The method demonstrated sensitivity to phase distortions of the order of $\lambda/10^4$, offering a valuable tool for characterizing weak cumulative nonlinear effects in photonic materials [1].

Nd^{3+}–Y^{3+}-Codoped CaF_2 and SrF_2 Crystals Under Femtosecond Excitation at 800 nm

Femtosecond Z-scan experiments on Nd^{3+}/Y^{3+}-codoped fluoride crystals (CaF_2 and SrF_2) at 800 nm indicate that although neodymium doping induces a measurable change in n_2, its magnitude is relatively small compared to other dopants. However, these experiments confirm that doping alters the third-order susceptibility of the host crystals, a clear demonstration of dopant-induced nonlinear refractive effects [2].

Yb^{3+}-Doped Double-Tungstate Crystals ($KRE(WO_4)_2$)

In monoclinic double tungstates doped with Yb^{3+} (where RE = Gd, Y, Yb, Lu), anisotropic nonlinear refractive indices (n_2) were measured by Z-scan The results demonstrated polarization-dependent n_2 values and showed a trend correlating the magnitude of n_2 to the ionic radius of the RE^{3+} ion [21]. These findings underline how rare-earth identity and crystal anisotropy influence Kerr-type nonlinear responses.

8.4 Applications of Population Lens Effects in Photonic Devices

The population lens effect, which arises from excited-state redistribution under intense optical excitation, offers promising avenues for the development of advanced photonic devices [22, 23]. In REI-doped systems, this nonlinear refractive index change is driven primarily by the dynamic variation of population levels within the electronic states of 4f [24]. As a consequence, the material behaves as a self-induced lens, whose focusing or defocusing character depends on the spatial distribution of excited ions.

This controllable refractive index modulation opens opportunities for passive and active optical control mechanisms in a variety of applications.

One of the most compelling applications is in optical limiting, where the material protects sensitive optical sensors or human eyes from damage by high-intensity laser pulses. Materials exhibiting population lensing can attenuate transmitted light at high fluences due to beam defocusing, thus reducing the risk of optical overload.

This effect is especially useful in systems operating under continuous-wave or long-pulse excitation, where the population build-up is significant.

In the realm of all-optical switching and modulation, population lensing provides a mechanism to induce intensity-dependent phase shifts that can be harnessed to control the propagation of optical signals. For instance, in interferometric setups such as Mach–Zehnder or Fabry–Pérot cavities, the nonlinear phase delay induced by population redistribution can be used to modulate the output signal purely by varying the input power [18]. This allows for high-speed, low-power optical logic elements compatible with integrated photonic circuits.

Another application is in beam shaping and self-focusing compensation. In optical systems where nonlinear self-focusing may lead to beam collapse or filamentation, a material with a well-characterized negative nonlinear refractive index due to population lensing can act as a compensating element. By tuning the excitation parameters or dopant concentration, one can engineer a nonlinear response to stabilize beam propagation in optical fibers, amplifiers, or high-power laser systems.

Furthermore, the spatiotemporal characteristics of the population lens effect offer avenues for time-resolved optical gating and imaging. Using the delayed refractive response relative to the excitation pulse, researchers have proposed dynamic wavefront shaping and lensing systems whose optical properties evolve over microsecond timescales. This is especially attractive for pump–probe spectroscopy, time-domain holography, and adaptive optics.

In emerging research, there is also growing interest in using population lens effects in reconfigurable photonic materials [25, 26], such as rare-earth-doped metasurfaces or hybrid systems combining REI with plasmonic or polymeric matrices. In these contexts, the nonlinear refractive response can be engineered to enable spatially and temporally tunable photonic functionalities, including diffraction control, optical memory, and signal routing.

In general, the population lens effect is not only a fundamental phenomenon explored by Z-scan but also a powerful mechanism that allows the design of smart optical materials and responsive photonic systems [13, 20]. As research advances in rare-earth-doped materials and their nonlinear properties, new architectures and applications based on population-induced refractive effects are expected to emerge, bridging the gap between nonlinear optics and practical photonic device engineering.

Therefore, nonlinear optical techniques, particularly the Z-scan method, have proven to be powerful tools for investigating third-order nonlinearities in REI-doped materials [8, 11, 18]. These techniques offer critical insights into the intensity-dependent refractive and absorptive properties of photonic materials, complementing conventional linear optical measurements. Among the various nonlinear effects, the population lens mechanism—originating from the redistribution of ions between ground and excited states under laser irradiation—has emerged as a fundamental contributor to the nonlinear refractive index in REI-doped systems.

The Z-scan technique, especially in its time-resolved variant, enables the distinction between instantaneous (electronic) and delayed (population- or thermal-based) contributions to the nonlinear response. This capability is particularly important for rare-earth-doped materials, where excited-state lifetimes are long enough to support temporally cumulative nonlinear effects. The use of normalized transmittance has proven to be highly effective in isolating these slow processes and eliminating linear artifacts.

Experimental results from a variety of REI-doped materials, such as Nd^{3+}- and Yb^{3+}-doped crystals, glasses, and nanostructures, reinforce the relevance of Z-scan analysis [2, 12]. For instance, time-resolved cw Z-scan experiments with Nd^{3+}-doped glasses revealed a pronounced population lens effect, in contrast to the fast Kerr-type response in ruby crystals. Likewise, in Yb^{3+}-doped tungstates, polarization-dependent measurements highlighted how host crystal symmetry and rare-earth identity influence the nonlinear response. Studies on $NaYF_4$:Yb, Er up-converting nanocrystals further illustrated the interplay between excitation intensity, thermal effects, and excited-state absorption in shaping the observed nonlinearity.

Altogether, these findings demonstrate that nonlinear refractive behavior in REI-doped materials is dependent not only on the ion species and host matrix but also on experimental parameters such as excitation intensity, wavelength, and pulse duration. The Z-scan technique stands out for its simplicity, sensitivity, and versatility, allowing the extraction of the real and imaginary components of third-order nonlinear susceptibility from a single experiment [10, 18, 27].

As photonic technologies continue to evolve toward more compact, efficient, and functionally diverse platforms, the ability to accurately characterize and understand nonlinear phenomena becomes increasingly important. The population lens effect, in particular, offers promising avenues for the development of all-optical switching, modulation, and limiting devices based on REI-doped systems [28]. Therefore, mastering Z-scan methodology and its theoretical underpinnings is indispensable for researchers aiming to design and optimize next-generation nonlinear optical materials and devices.

References

1. T. Xia, D.J. Hagan, M. Sheik-Bahae, E.W. Van Stryland, Opt. Lett. **19**, 317 (1994)
2. Y. Guo, S. Lu, L. Su, C. Zhao, H. Zhang, S. Wen, Appl. Opt. **54**, 953 (2015)
3. M. Sheik-Bahae, A.A. Said, T.-H. Wei, D.J. Hagan, E.W. Van Stryland, IEEE J. Quantum Electron. **26**, 760 (1990)
4. V.S. Zanuto, J.F.M. Dos Santos, M.L. Baesso, T. Catunda, Opt. Laser Technol. **142**, 107248 (2021)
5. J. Wang, M. Sheik-Bahae, A.A. Said, D.J. Hagan, E.W. Van Stryland, J. Opt. Soc. Am. B **11**, 1009 (1994)
6. C. Cassagne, O. Ba, G. Boudebs, Materials **15**, 5008 (2022)
7. A.M. Peter, M. Ramya, M. Kailasnath, J. Phys. Conf. Ser. **2357**, 012013 (2022)
8. A.G. Pelosi, F.R. Henrique, J.M.P. De Almeida, M. Nalin, D.F. Franco, L.H.Z. Cocca, C.R. Mendonça, in *Latin America Optics and Photonics (LAOP) Conference 2022* (Optica Publishing Group, Recife, 2022), p. Tu4A.56.
9. L. Giordano, D.L. Fritzen, J.M. Gonçalves, L.C.V. Rodrigues, J.M.A. Caiut, in *Modern Luminescence from Fundamental Concepts to Materials and Applications, Volume 2* (Elsevier, 2025), pp. 101–132.
10. A. Ajami, A. Ovsianikov, R. Liska, S. Baudis, Appl. Phys. B **130**, 138 (2024)
11. N.S. Camilo, J.C.S. Filho, V. Pilla, D.N. Messias, A.C.A. Silva, N.O. Dantas, A.A. Andrade, J. Non-Cryst. Solids **612**, 122345 (2023)
12. A. Volpi, J. Kock, A.R. Albrecht, M.P. Hehlen, R.I. Epstein, M. Sheik-Bahae, Opt. Lett. **46**, 1421 (2021)
13. A.K. Kar, in *13th International Conference on Fiber Optics and Photonics* (OSA, Kharagpur, 2016), p. Tu3D.2
14. V.S. Ganesha Krishna, M.G. Mahesha, Physica B: Condensed Matter **628**, 413628 (2022)
15. D.M. Marom, in *Comprehensive Microsystems* (Elsevier, 2008), pp. 219–265
16. J.F.M. Dos Santos, G.G.G. Costa, D.N. Messias, T. Catunda, Opt. Commun. **479**, 126421 (2021)
17. S. Kostritskii, M. Aillerie, E. Kokanyan, O. Sevostyanov, Opt. Quant. Electron. **52**, 92 (2020)
18. E.W. Van Stryland, M. Sheik-Bahae, in (Bellingham, United States, 1997), p. 102910Q
19. M. Kilinc, U. Demirbas, J.B. Gonzalez-Diaz, J. Thesinga, M. Kellert, G. Palmer, F.X. Kärtner, M. Pergament, Opt. Mater. Express **13**, 3200 (2023)
20. N. Ghazyani, M.H. Majles Ara, M. Raoufi, RSC Adv. **10**, 25696 (2020)
21. N. Thilmann, G. Strömqvist, M.C. Pujol, V. Pasiskevicius, V. Petrov, F. Díaz, Appl. Phys. B **96**, 385 (2009)

22. A.H. Almuqrin, M.I. Sayyed, A.G. Pramod, K. Keshavamurthy, G. Jagannath, J. Mater. Sci. Mater. Electron. **33**, 16357 (2022)
23. X. Yang, Q. Li, X. Meng, D. Li, ACS Omega **5**, 15821 (2020)
24. S. Liu, D. Serrano, A. Fossati, A. Tallaire, A. Ferrier, P. Goldner, RSC Adv. **8**, 37098 (2018)
25. J.H. Ko, Y.J. Yoo, Y. Lee, H.-H. Jeong, Y.M. Song, iScience **25**, 104727 (2022)
26. C. Li, R. Pan, C. Gu, H. Guo, J. Li, Advanced Science **11**, 2306344 (2024)
27. N. Passilly, E. Haouas, V. Ménard, R. Moncorgé, K. Aı¨t-Ameur, Opt. Commun. **260**, 703 (2006)
28. B. Mitchell, H. Austin, D. Timmerman, V. Dierolf, Y. Fujiwara, Nanophotonics **10**, 851 (2020)

Chapter 9
Judd–Ofelt Analysis

Rare-earth ions (REI) exhibit the $4f^n 5s^2 5p^6$ electron configuration, where n goes from 1 to 13, and the REI displays narrow and relatively intense spectral lines, which occur in transitions between the levels of the $4f$ orbital via electric dipole (ED). Nevertheless, these $4f \rightarrow 4f$ transitions are not allowed[1] by the parity selection rules, i.e., ED transitions occur between states of opposite parity (even to odd and vice versa). Further, only magnetic-dipole (MD) and electric quadrupole transitions are possible in within the $4f$ orbital, but with low probabilities, which should be reflected in weak spectral lines. Hence, the origin of the REI spectral intensities should be mainly due to ED contribution and the selection rules should be modified. In 1962, Judd [1] and Ofelt [2] independently developed, a theoretical model to describe the transitions between the $4f$ levels of REI, and it is known as Judd–Ofelt (JO) theory. They consider those transitions are allowed due to the mixture of $4f$ and $5d$ states with opposite parity. Besides, when a REI is hosted in a solid, they experiment with a crystal-field potential (V_{CF}), which acts as a perturbative potential and distorts the electronic motion of the $4f$ electrons. In this manner, the selection rules are relaxed, and the observed transitions become allowed by induced and forced ED transitions [3].

The JO theory satisfactorily explains the spectral intensities for most of the REI by calculating three intensity parameters, known as the JO parameters, which are sensitive to the local environment where the REI is located. In addition, these JO parameters provide information about the structural properties of the luminescent material per se, as well as the bonding and symmetries around the REI [4]. This is a semi-empirical way to study the structural properties of REI-doped glasses and to predict their lasing performance [5] based, traditionally, on their absorption spectrum. Regarding this last point, there are studies that use the emission spectrum to calculate the JO parameters for Eu^{3+} since, usually, high energy levels are not observed because their energies go beyond the glass absorption bandgap [6]. Another REI with a

[1] Not allowed or forbidden means that these transitions occur with low probability.

G. Lozano C. et al., *A Brief Guide to Optical Spectroscopy of Trivalent Rare Earth Ions in Glasses*, Springer Series in Optical Sciences 247,
https://doi.org/10.1007/978-3-032-18501-3_9

particular JO analysis is the Pr^{3+} because, in some cases, it is reported negative JO parameters, and modified JO models are proposed [7–10]. The reasons behind this shall be discussed later. This chapter is focused on the standard JO analysis based on the absorption spectra.

9.1 Fundamentals

When a REI is within a medium, e.g., crystal and glass, it experiments different kinds of interactions as potential energy, and its Hamiltonian is expressed as

$$H = -\frac{\hbar^2}{2m}\sum_{j}^{N}\nabla_j^2 - \sum_{j}^{N}\frac{Ze^2}{r_j} + \sum_{j<k}^{N}\frac{e^2}{r_{j,k}} + \sum_{j}^{N}\xi(r_j)(S_j\cdot L_j) + V_{CF} \tag{9.1}$$

where the first term is the kinetic energy of the $4f$ electrons, the second term is the coulomb interaction between the nucleus and the $4f$ electrons, the third term is the electron–electron electrostatic interaction, the fourth term is the spin–orbit (Russel–Saunders) coupling Hamiltonian, where $\xi(r_j)$ is the coupling function [11]. These first four terms are the Hamiltonian of a free REI. Besides, the REI interacts with the electrostatic field because of the distribution charge within the environment. This electrostatic field is the crystal-field (CF) potential, denoted as V_{CF}, and distorts the $4f$ orbital. Each kind of interaction generates a splitting on the $4f^n$ levels: The electron–electron repulsion splits the degenerate levels, characterized by the total spin quantum number S and by the angular momentum L, being defined by the spectroscopic notation ^{2S+1}L. (ii) The spin–orbit coupling splits into $^{2S+1}L_J$ levels, where $J = S + L$. (iii) Finally, the CF potential divide the $^{2S+1}L_J$ into $2J + 1$ levels (Stark levels).

Along this guide, we analyzed several REI transitions, for instance, the Er^{3+}: $^4I_{13/2} \rightarrow {}^4I_{15/2}$, which corresponds to the wavelength $\approx$ 1530 nm (6536 cm^{-1}), and is mainly employed in optical communications since it covers the S, C and L communication windows [12]. As can be seen, this transition does not obey the Laporte selection rules, i.e., $\Delta L = \pm 1$, $\Delta S = 0$, and $\Delta J = \{0, \pm 1\}$ ($0 \nleftrightarrow 0$). Since the REI is hosted in a non-symmetric lattice, due to the non-immersion symmetry of the V_{CF}, the selection rules are relaxed, and the observed transitions become allowed by induced and forced ED transitions [3]. The JO theory considers that these transitions are allowed due to the mixture of the $4f$ (even parity) and $5d$ (opposite parity) states and obtained an expression of the ED line strength $S_{ED}(J, J')$, which is related to the transition probabilities. Furthermore, some assumptions are made to develop their model and simplify their calculations. First, the V_{CF} is considered as a first-order perturbation. Further, the Hamiltonian spin–orbit coupling is of the same order of magnitude as the electron–electron electrostatic interaction, which is known as intermediate coupling. Besides, the energy of the levels above the $4f$ configuration are degenerated, and the excited states are positioned much higher than the ground one [13]. After some algebra, Judd and Ofelt obtained the following expression:

$$S_{ED}\left(J,J^{'}\right)=e^{2}\sum_{\lambda}\Omega_{\lambda}\left|\langle S,L,J|U^{\lambda}\left|S^{'},L^{'},J^{'}\rangle\right.\right|^{2} \quad (9.2)$$

where e the fundamental electron charge (4.803×10^{-10} $cm^{3/2}\cdot g^{1/2}\cdot s^{-1}$), Ω_λ ($\lambda = 2, 4, 6$) are the JO intensity parameters, and U^λ is a unit tensor, being $|\langle S,L,J|U^{\lambda}|S',L',J'\rangle|^{2}$ the elements of a reduced matrix. These elements were calculated by Carnall et al. [14], and are independent on the kind of host matrix where the REI is located. With the JO parameters we can predict features about bonding and the local structure in the vicinity of REI, as well as the degree of symmetry in the REI sites. In this case, Ω_2 is related to the symmetry and structural changes, while Ω_4 and Ω_6 are related to the rigidity, dielectric properties and viscosity of the host material [15].[2] It is worth noting that the selection rules for ED transitions of REI derived from the JO theory are $\Delta S = 0$ and ΔL, $\Delta J \leq \pm 6$ ($0 \nleftrightarrow 0$ for is worth noting worth noting that the selection REI ED transitions of REI derived from the JO theory are $\Delta S = 0$ and ΔL, $\Delta J \leq \pm 6$ ($0 \nleftrightarrow 0$ for J) [17].

Furthermore, the JO parameters are useful to determine other spectroscopic parameters, i.e., the spontaneous emission probabilities, radiative lifetime, and branching ratios. First, we should consider the magnetic-dipole (MD) transitions, with selection rules $\Delta S = \Delta L = 0$, $\Delta J \leq \pm 1 (0 \nleftrightarrow 0)$, and the MD line strength between two states, which is given by

$$S_{MD}\left(J,J^{'}\right)=\left(\frac{eh}{2\pi m_{e}c}\right)^{2}\left|\langle S,L,J|\hat{L}+2\hat{S}\left|S^{'},L^{'},J^{'}\rangle\right.\right|^{2} \quad (9.3)$$

Here, h is Planck's constant (6.626×10^{-27} $g\cdot cm^{2}\cdot s^{-1}$), m_e is the electron mass (9.109×10^{-28} g), and c is the speed of light at vacuum ($\approx 2.9979 \times 10^{10}$ $cm\cdot s^{-1}$). The matrix elements of the MD operator for cases allowed by $\Delta J \leq \pm 1$ are tabulated in Table 9.1, and only depend on the quantum numbers of the initial and final states.

The oscillator strength is a quantity that is proportional to the probability that an ED transition occurs. The theoretical or calculated oscillator strength, denoted as f_{cal}, for an ED transition from the ground state is [18]

$$f_{cal}=\frac{8\pi^{2}m_{e}c}{3he^{2}\lambda(2J+1)}\frac{\left(n^{2}+2\right)}{9n}S_{ED} \quad (9.4)$$

[2] The parameter Ω_2 used to be associated to covalency, however, a reinterpretation of the JO parameters suggests that covalency is more related to Ω_4 and Ω_6 and Ω_2. The authors highly recommend the Ref. [16], where it is discussed the interpretation of the JO parameters in solids, gases, and complexes.

Table 9.1 MD operator allowed for each particular $\Delta J \leq \pm 1$ case

$\Delta J \leq \pm 1$	$\langle S, L, J\|\hat{L} + 2\hat{S}\|S', L', J'\rangle$
$J = J'$	$\left[\frac{1+J(J+1)+S(S+1)-L(L-1)}{2J(J+1)}\right][J(J+1)(2J+1)]^{1/2}$
$J = J' - 1$	$\left[\frac{(S+L+J+1)(S+L-J+1)(S+J-L)(L+J-S)}{4J}\right]^{1/2}$
$J = J' + 1$	$\left[\frac{(S+L+J+2)(S-L+J+1)(L-S+J+1)(L+S-J)}{4J(J+1)}\right]^{1/2}$

where λ is the wavelength of the electronic transition and n is the refractive index of the medium. Additionally, the spontaneous emission probability considering both electric and magnetic-dipole transitions is

$$A\left(J, J'\right) = \frac{64\pi^4}{3h\lambda^3(2J+1)}\left[\frac{\left(n^2+2\right)}{9n}S_{\text{ED}} + n^3 S_{\text{MD}}\right] \tag{9.5}$$

Then, the radiative calculated lifetime can be obtained from the spontaneous emission probability as

$$\tau_{\text{cal}} = \left[\sum_{J'} A\left(J, J'\right)\right]^{-1} \tag{9.6}$$

In this case, the sum is over all transition probabilities for the levels above the specific emission state. Another spectroscopic parameter that can be derived is the luminescence branching ratio β, which characterizes the probability of exciting stimulated emission in a specific channel [19]. This parameter is calculated as

$$\beta = \tau_{\text{cal}} A\left(J, J'\right) \tag{9.7}$$

9.2 Oscillator Strength Calculation

In the last section, we explained how to calculate the oscillator strength with Eq. (9.4). Nevertheless, we can calculate the experimental oscillator strength f_{exp} (unitless), which is proportional to the area under the absorption band defined as [18]

$$f_{\text{exp}} = \frac{m_e c}{\pi e^2}\int \sigma_{\text{abs}}(\nu)d\nu \tag{9.8}$$

where $\sigma_{abs}(\nu)$ is the absorption cross-section in function of the frequency ν (in s^{-1}). Hence, first we need to measure the thickness of the sample density, and then the absorption spectra. In this case, the raw materials for the prepared samples are TeO_2, ZnO, and GeO_2, and four samples were fabricated varying the Er^{3+} concentration by as 0.25, 0.50, 0.75, and 1.00 mol%, maintaining fixed at 3 mol% of Yb^{3+}, and were labeled according to the Er^{3+} concentration (0.25Er, 0.50Er, 0.75Er, and 1.00Er). Besides, the thickness t of both samples is 1.5 mm, the density ρ is 5.63, 5.64, 5.70, and 5.73 g·cm^{-3} for samples 0.25Er, 0.50Er, 0.75Er, and 1.00Er, respectively. We remark that these glasses are used in Chap. 5 for optical thermometry, and the addition of Yb^{3+} is justified.

Figure 9.1a illustrates the absorption spectra of both glasses (with no baseline correction), and the bands centered at 378, 407, 443, 450, 488, 522, 545, 801, and 976 nm corresponds to the electronic transitions from the Er^{3+} ground state $^4I_{15/2}$ to the excited states $^4G_{11/2}$, $^2H_{9/2}$, $^4F_{3/2}$, $^4F_{5/2}$, $^4F_{7/2}$, $^2H_{11/2}$, $^4S_{3/2}$, $^4F_{9/2}$, $^4I_{9/2}$, and $^4I_{11/2}$, respectively, whereas the large band centered at ≈ 976 nm is mostly attributed to the Yb^{3+} transition $^2F_{7/2} \rightarrow {}^2F_{5/2}$ due to its high concentration. Moreover, there are two characteristic peaks of the Er^{3+} in these glasses, i.e., $^4I_{15/2} \rightarrow {}^4G_{11/2}$ and $^4I_{15/2} \rightarrow {}^2H_{11/2}$, which are known as hypersensitive transitions,[3] and exhibit the most intense intensities in the spectra. This kind of transition shall be discussed later.

Applying the expression to calculate the number of Er^{3+} ions per volume $N = 2N_a x_i \rho/\mathcal{M}$, where N_a is the Avogadro's number (6.022×10^{23} mol^{-1}), x_i is the Er^{3+} mol concentration, and $\mathcal{M}$ is the molar mass of the sample, the $\sigma_{abs}(\nu)$ was obtained for each sample and it is displayed in Fig. 9.1b, where the spectra were vertically shifted to distinguish the absorption bands. It is pertinent to highlight that the spectral region in Fig. 9.1b only considers the Er^{3+} absorption bands. Removing the baseline is useful for the visibility and accuracy of the analysis of the peaks of interest. The baseline is the spectral region with, in our case, no peaks, and, as we can see in Fig. 9.1b, the baseline is not perfectly horizontal. For this reason, it is not just subtracting the minimum value of the spectrum to shift it to zero. Therefore, data points should be selected to interpolate them in a manner that the baseline is generated. Such data points can be calculated with different methods, e.g., second derivatives (zeroes) or first and second derivatives (zeroes). However, due to the complexity of the spectrum, these data points were selected by user criteria and connected with a spline interpolation method.

Once the baseline is calculated, we can proceed with the correction as seen in Fig. 9.1c, where 10 absorption bands were chosen for the oscillator strengths calculations. Herein, the spectra were vertically shifted to facilitate the identification of the peaks in each sample, where it is observed that the peak position does not change for both samples. Afterward, the area of each absorption band was computed by numerical integration, and their values are plotted in Fig. 9.1d, where the maximum values are ascribed to the electronic transitions of $^4I_{15/2} \rightarrow {}^4G_{11/2}$ and $^4I_{15/2} \rightarrow {}^2H_{11/2}$

[3] Hypersensitive transitions obey the selection rule $|\Delta J| \leq 2, |\Delta L| \leq 2\, and\, \Delta S = 0$, however, the $^4I_{15/2} \rightarrow {}^2H_{11/2}$ is considered hypersensitive, although it does not obey this rule [20].

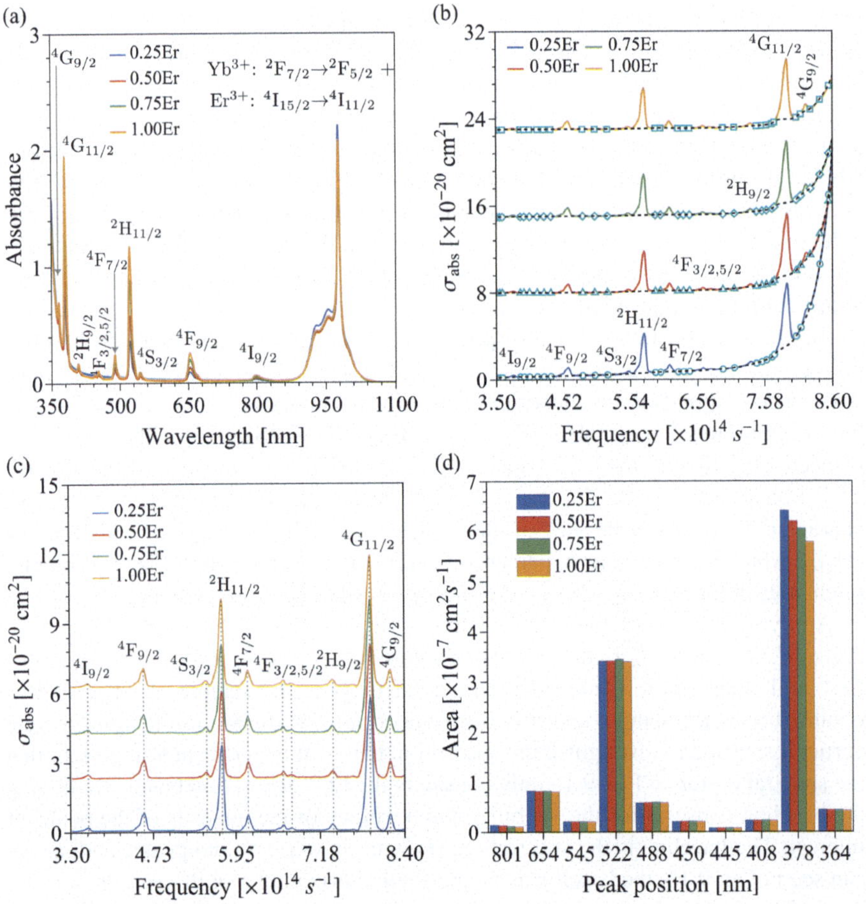

Fig. 9.1 **a** Absorption spectra of the Er^{3+}-Yb^{3+}-doped glasses, revealing the characteristic Er^{3+} and Yb^{3+} absorption bands. **b** Cross-section absorption of the samples with their respective baselines (dashed lines). The plot markers indicate the selected data points to create the baseline. **c** Absorption cross-section with baseline correction. The spectra were vertically shifted for reading purposes. **d** Area of the selected absorption bands from **c** for the oscillator strengths calculations

electronic transitions. Hence, these transitions are expected to display the largest oscillator strengths in the next calculations.

9.3 Judd–Ofelt Parameters Calculation

Equations (9.4) and (9.8) are essentially the same, and we can compare both to obtain the JO parameters as

$$S_{ED}^{\exp} = \frac{9n}{\left(n^2+2\right)^2}\frac{3h\lambda(2J+1)}{8\pi^2 m_e c} f_{\exp} = e^2\sum_{\lambda}\Omega_\lambda \left|\langle S, L, J|U^\lambda|S', L', J'\rangle\right|^2 \quad (9.9)$$

where $S_{ED}^{\exp}$ is the experimental ED line strength. This last expression can be written in matrix representation for M transitions as

$$S_{ED}^{\exp} = \begin{pmatrix} \|U_1^2\|^2 & \|U_1^4\|^2 & \|U_1^6\|^2 \\ \|U_2^2\|^2 & \|U_2^4\|^2 & \|U_2^6\|^2 \\ \vdots & \vdots & \vdots \\ \|U_M^2\|^2 & \|U_M^4\|^2 & \|U_M^6\| \end{pmatrix} \begin{pmatrix} \Omega_2 \\ \Omega_4 \\ \Omega_6 \end{pmatrix} \quad (9.10)$$

Here, we are dealing with an overdetermined system, since the number of unknowns is lower than the number of equations. Consequently, the standard least-squares method must be applied to minimize the absolute differences between the theoretical and experimental values [21]. Equation (9.10) can be expressed in a simplified form as $S = U\Omega$, being S a M × 1 matrix for the $S_{ED}^{\exp}$ elements, U is a M × 3 matrix for the matrix elements of $\left|\langle S, L, J|U^\lambda|S', L', J'\rangle\right|^2$, and Ω is the 3 × 1 matrix for the JO intensity parameters. Applying the least-square method, the matrix Ω is

$$\Omega = \left(U^+U\right)^{-1}U^+S \quad (9.11)$$

To validate the calculation of the JO parameters, the root-mean square deviation (δ_{rms}) between the calculated and experimental ED line strengths for the M transitions is $\delta_{rms} = [\sum\left(S_{ED}^{exp} - S_{ED}^{cal}\right)^2/(M-3)]^{1/2}$, where S_{ED}^{cal} is given by Eq. (9.4). Besides, the δ_{rms} for the oscillator strength calculation is $\delta_{rms} = [\sum\left(f_{\exp} - f_{\mathrm{cal}}\right)^2/(M-3)]^{1/2}$.

At this moment, the last quantity we need to compute the JO parameters is the refractive index. In Table 9.2 is tabulated the wavelength-dependent refractive indices of the samples (measured with the prism method) and the parameters A and B of the two-term Sellmeier's dispersion relation (see Chap. 7). For the calculation of JO parameters, it is assumed that the fit of $n(\lambda)$ can be extrapolated until the lowest selected wavelength from Fig. 9.1, i.e., 364 nm. This assumption was made because the minimum wavelength of the experimental setup is 532 nm, and we need to cover the most relevant absorption bands of the Er^{3+}, such as the hypersensitive transitions. Therefore, having a large enough number of absorption bands (at least five or six) brings more accurate results.

At this stage, we can determine the JO parameters, and the results shall be briefly discussed. First, we need to set our equations system from Eq. (9.10) with the help of the elements of a reduced matrix given by Carnall et al. [14], and these elements for others REI are in Appendix. For the Er^{3+} and using the $S_{ED}^{\exp}$ (in cm^2) of the sample 0.25Er, we have the following equations system:

Table 9.2 Refractive indices of the Er and Er-Yb samples at a specific wavelength λ (in nm) and the Sellmeier parameters *A* and *B* ($R^2 > 0.99$)

λ	$n(\lambda)$			
	0.25Er	0.50Er	0.75Er	1.00Er
532	2.056	2.052	2.051	2.048
633	2.033	2.026	2.025	2.023
973	1.996	1.994	1.993	1.991
1308	1.986	1.983	1.982	1.980
1538	1.980	1.977	1.976	1.975
A	2.889	2.878	2.876	2.869
B	30,058	29,461	29,256	28,968

$$\begin{array}{lllllll}
0\Omega_2 & + & 0.1856\Omega_4 & + & 0.0122\Omega_6 & = & 2.89\times10^{-21} \\
0\Omega_2 & + & 0.5275\Omega_4 & + & 0.4612\Omega_6 & = & 1.46\times10^{-20} \\
0\Omega_2 & + & 0\Omega_4 & + & 0.223\Omega_6 & = & 3.07\times10^{-21} \\
0.7326\Omega_2 & + & 0.4222\Omega_4 & + & 0.0927\Omega_6 & = & 4.69\times10^{-20} \\
0\Omega_2 & + & 0.1467\Omega_4 & + & 0.628\Omega_6 & = & 7.38\times10^{-21} \\
0\Omega_2 & + & 0\Omega_4 & + & 0.2237\Omega_6 & = & 2.41\times10^{-21} \\
0\Omega_2 & + & 0\Omega_4 & + & 0.1256\Omega_6 & = & 8.99\times10^{-22} \\
0\Omega_2 & + & 0.0157\Omega_4 & + & 0.2278\Omega_6 & = & 2.34\times10^{-21} \\
0.897\Omega_2 & + & 0.5123\Omega_4 & + & 0.172\Omega_6 & = & 6.92\times10^{-20} \\
0.0229\Omega_2 & + & 0.0043\Omega_4 & + & 0.0794\Omega_6 & = & 3.94\times10^{-21}
\end{array}$$

In Table 9.3 are listed the experimental and theoretical oscillator strengths for all glasses and the root mean square deviations δ_{rms}. The small values of δ_{rms} indicate a good agreement between f_{exp} and f_{cal}. Besides, the $^4I_{15/2} \rightarrow {}^4G_{11/2}$ absorption band exhibits the highest oscillator strength values, as well as the $^4I_{15/2} \rightarrow {}^2H_{11/2}$ transition, being approximately half of those obtained for 378 nm. The f_{exp} values of the $^4I_{15/2} \rightarrow {}^2H_{11/2}$ band of theses glasses are greater than reported in other tellurite glasses [5, 20, 22]. Regarding the JO intensity parameters, Ω_2 exhibit a slight increment with the addition of Er^{3+}, which indicates an increment of the asymmetry around the Er^{3+} ions. This increment is associated with the disorder in the glass host. A subtle increment in Ω_6 is also observed with the addition of Er^{3+}, conversely Ω_4, and suggest an increment on the viscosity and rigidity of the glass matrix [23]. Indeed, it is common to observe increments in viscosity by adding oxides with high melting points, in this case, the Er_2O_3 (melting point at approximately 2420 °C [24]). Increments in Ω_4 are associated to an increase in the covalence between O^{2-} ions and the Er^{3+} [25]. Further, quantity Ω_4/Ω_6 is known as the spectroscopic quality factor and predicts the magnitude of stimulated emission in a laser active medium [26]. The Ω_4/Ω_6 values are higher than those reported elsewhere [26–28], which indicates relative intense laser emission dynamics. An additional characteristic of the obtained JO parameters is that they follow the trend $\Omega_6 < \Omega_4 < \Omega_2$, which is also reported in other Er^{3+}/Er^{3+}-Yb^{3+}-doped tellurite glasses [27, 29, 30].

The standard JO analysis can be successfully performed in most of the REI, e.g., Tm^{3+} [31], Sm^{3+} [30], Dy^{3+} [32], Ho^{3+} [33], Tb^{3+} [34], and Nd^{3+} [35]. In the case of Ce^{3+} and Yb^{3+}, it is not possible to apply the JO theory since these REI only exhibit one electronic transition. Moreover, applying the original JO theory in Pr^{3+}, this leads to negative Ω_2 and lifetime values, which has no physical meaning. This last occurs when the Pr^{3+}: $^3H_4 \rightarrow {}^3P_2$ hypersensitive transition (centered at ≈443 nm [12]) is included in the calculations [36]: The $^3P_2$ level is close to the ground state of the 5*d* configuration, and the JO theory assumes that this energy gap is large. In some research, the Pr^{3+}: $^3H_4 \rightarrow {}^3P_2$ absorption band is omitted, nonetheless, the small gap of the 4*f*–5*d* configuration affects the ED transitions [37]. In this framework, different modified JO models were proposed, for instance, the developed by Kornienko et al., which considers the dependence of the ED line strength with the energy of the 5d states [38]. For further discussion about the JO analysis of the Pr^{3+}, the reader is referred to Refs. [37, 39–41]. Another REI with special energy level structure is the Eu^{3+}, where the observed electronic transitions exhibit weak absorption intensities in the visible region [42]. In this case, the JO parameters are calculated using the emission spectrum of the Eu^{3+} using the FIR of the Eu^{3+}: $^5D_0 \rightarrow {}^7F_{\lambda=2,4,6}$ (ED transitions) and Eu^{3+}: $^5D_0 \rightarrow {}^7F_0$ (MD transition) emission bands [43], where it is

Table 9.3 Experimental and calculated oscillator strengths ($\times 10^{-6}$) each absorption band of the samples. The δ_{rms} values for the oscillator strengths calculation are given in units of $\times 10^{-6}$. The Judd–Ofelt parameters are in $\times 10^{-20}$ cm^2

Abs. band	λ_{abs}	0.25Er		0.50Er		0.75Er		1.00Er	
		f_{exp}	f_{cal}	f_{exp}	f_{cal}	f_{exp}	f_{cal}	f_{exp}	f_{cal}
$^4I_{15/2} \rightarrow {}^2G_{9/2}$	364	1.69	0.87	1.64	0.86	1.64	0.86	1.61	0.83
$^4I_{15/2} \rightarrow {}^4G_{11/2}$	378	23.9	23.4	23.4	23.4	22.8	23.1	21.8	22.4
$^4I_{15/2} \rightarrow {}^2H_{9/2}$	408	0.87	0.88	0.87	0.88	0.88	0.90	0.86	0.89
$^4I_{15/2} \rightarrow {}^4F_{3/2}$	445	0.29	0.38	0.29	0.38	0.29	0.39	0.29	0.39
$^4I_{15/2} \rightarrow {}^4F_{5/2}$	450	0.78	0.66	0.79	0.67	0.79	0.68	0.78	0.68
$^4I_{15/2} \rightarrow {}^4F_{7/2}$	488	2.18	2.50	2.19	2.49	2.23	2.53	2.20	2.49
$^4I_{15/2} \rightarrow {}^2H_{11/2}$	522	12.8	13.1	12.9	12.8	12.9	12.7	12.8	12.3
$^4I_{15/2} \rightarrow {}^4S_{3/2}$	545	0.79	0.53	0.78	0.53	0.79	0.55	0.78	0.54
$^4I_{15/2} \rightarrow {}^4F_{9/2}$	654	3.10	2.98	3.06	2.95	3.07	2.95	3.03	2.89
$^4I_{15/2} \rightarrow {}^4I_{9/2}$	801	0.49	0.61	0.49	0.60	0.45	0.59	0.43	0.58
δ_{rms}		0.38		0.34		0.38		0.44	
Ω_2		5.34		5.26		5.21		5.05	
Ω_4		1.86		1.84		1.82		1.78	
Ω_6		0.91		0.92		0.95		0.94	
Ω_4/Ω_6		2.04		1.99		1.92		1.90	
δ_{rms}		0.11		0.09		0.11		0.13	

typically reported the values of the Ω_2 and Ω_4 parameters. The absence of the Ω_6 parameter is due to the lack of the Eu^{3+}: $^5D_0 \rightarrow {}^7F_6$ transition [44–46].

The calculation of the radiative spectroscopic parameters derived from the JO theory not only involves transitions from an excited state to the ground state. Conventional JO analysis requires information from transitions between intermediate energy levels, as seen in Eq. (9.6) for the calculated radiative lifetime estimation. Thereby, we need to calculate the S_{ED} and S_{MD} to find the transition probabilities $A(J, J\prime)$, see Eq. (9.5), for all inter manifold $J \rightarrow J\prime$ transitions. For S_{MD}, their values can be computed using Eq. (9.3) and with the help of Table 9.1, whereas for S_{ED}, we need to use a squared reduced matrix, which considers all possible intermediate transitions. These elements for Er^{3+}, and other REI, can be found in Ref. [47]. Therefore, the spectroscopic parameters for some Er^{3+} transitions of samples 0.25Er and 1.00Er are given in Table 9.4. These bands were chosen because they exhibit some interesting features. First, in general, the branching ration of the transitions from an excited state to the ground state $^4I_{15/2}$ exhibits the largest value. The branching ratio of the $^4S_{3/2} \rightarrow {}^4I_{15/2}$ transition is about 67%, and this value is also observed in other tellurite glasses. Moreover, due to the selection rules of the MD transitions, few transitions are allowed, for instance, the $^2H_{9/2} \rightarrow {}^2H_{11/2}$, where the transition probability is 16 s^{-1} for both samples. In some cases, the MD transitions are disregarded due to their low line strength values, and they do not significantly contribute to the calculations of the transition probabilities. Other MD transitions are $^4F_{3/2} \rightarrow {}^2H_{9/2}$, $^4F_{5/2} \rightarrow {}^2H_{11/2}$, $^4F_{5/2} \rightarrow {}^4F_{7/2}$, $^4F_{7/2} \rightarrow {}^4F_{9/2}$ and $^4I_{9/2} \rightarrow {}^4I_{11/2}$.

As demonstrated, the JO parameters bring relevant information about structural properties of the glasses and the bonding between the REI and the ligand anions. The JO analysis has been successfully corroborated with the experiments during these years, which gives validity to this theory and a semi-empirical approach to structural properties.

9.4 Covalency Model of Hypersensitivity

Trivalent REIs have the 4*f* electrons shell that is shielded by the closed 5 s and 5p shells. As a result, the surrounding ligand environment has only a weak influence on the electronic cloud of the lanthanide ion. In this sequence, the covalency model of hypersensitivity, crystal-field theory, and the dynamic coupling mechanism together with Judd–Ofelt theory complement the spectroscopies study of trivalent REIs in glass materials. It is worth mentioning that the hypersensitive transitions violate the standard Laporte parity selection rule ($\Delta\ell = \pm 1$, where ℓ is the orbital angular momentum), which normally makes 4*f*–4*f* transitions very weak because they are parity-forbidden, where the oscillator strengths, are about constant, $f \approx 10^{-6}$ [48]. In this context, all these studies remain under active investigation, as the disordered structure of glass significantly influences 4*f*–4*f* electronic transitions. For instance, site-to-site variations can lead to shifts in electronic energy levels, variations in excited-state lifetimes (both short and long), differences in transition probabilities,

Table 9.4 Electric- and magnetic-dipole transition probabilities A_{ED} and A_{MD} (in s^{-1}), calculated lifetimes τ_{cal} (in ms) and branching ration β (in %) for some Er^{3+} transitions

Emission band	0.25Er			1.00Er		
	A_{ED}	A_{MD}	β	A_{ED}	A_{MD}	β
$^4G_{11/2} \rightarrow {}^4I_{15/2}$	70,550		89.6	65,268		89.4
$^4G_{11/2} \rightarrow {}^4I_{13/2}$	5477		6.9	5197		7.1
$^4G_{11/2} \rightarrow {}^4I_{11/2}$	236		0.3	225		0.3
$^4G_{11/2} \rightarrow {}^4I_{9/2}$	605		0.8	567		0.8
$^4G_{11/2} \rightarrow {}^4F_{9/2}$	1617		2.1	1510		2.1
$^4G_{11/2} \rightarrow {}^4S_{3/2}$	61		0.1	57		0.1
$^4G_{11/2} \rightarrow {}^2H_{11/2}$	60		0.1	57		0.1
$^4G_{11/2} \rightarrow {}^4F_{7/2}$	76		0.1	72		0.1
$^4G_{11/2} \rightarrow {}^4F_{5/2}$	5.9		0.0	5.8		0.0
$^4G_{11/2} \rightarrow {}^4F_{3/2}$	4.1		0.0	4.0		0.0
$^4G_{11/2} \rightarrow {}^2H_{9/2}$	6.7		0.0	6.5		0.0
τ_{cal}		0.01			0.01	
$^2H_{9/2} \rightarrow {}^4I_{15/2}$	2546	16	35.7	2535	16	36.7
$^2H_{9/2} \rightarrow {}^4I_{13/2}$	3489		48.9	3352		48.5
$^2H_{9/2} \rightarrow {}^4I_{11/2}$	851		11.9	812		11.7
$^2H_{9/2} \rightarrow {}^4I_{9/2}$	99		1.4	93		1.3
$^2H_{9/2} \rightarrow {}^4F_{9/2}$	55		0.8	53		0.8
$^2H_{9/2} \rightarrow {}^4S_{3/2}$	0.7		0.0	0.7		0.0
$^2H_{9/2} \rightarrow {}^2H_{11/2}$	46		0.9	43		0.9
$^2H_{9/2} \rightarrow {}^4F_{7/2}$	26		0.4	24		0.4
$^2H_{9/2} \rightarrow {}^4F_{5/2}$	0.9		0.0	0.9		0.0
$^2H_{9/2} \rightarrow {}^4F_{3/2}$	0.2		0.0	0.2		0.0
τ_{cal}		0.12			0.13	
$^2H_{11/2} \rightarrow {}^4I_{15/2}$	17,624		95.9	16,442		95.9
$^2H_{11/2} \rightarrow {}^4I_{13/2}$	283		1.5	269		1.6
$^2H_{11/2} \rightarrow {}^4I_{11/2}$	169		0.9	160		0.9
$^2H_{11/2} \rightarrow {}^4I_{9/2}$	229		1.2	218		1.3
$^2H_{11/2} \rightarrow {}^4F_{9/2}$	58		0.3	55		0.3
$^2H_{11/2} \rightarrow {}^4S_{3/2}$	0.1		0.0	0.1		0.0
τ_{cal}		0.05			0.06	
$^4S_{3/2} \rightarrow {}^4I_{15/2}$	1988		67.7	2011		67.7
$^4S_{3/2} \rightarrow {}^4I_{13/2}$	767		26.1	779		26.2
$^4S_{3/2} \rightarrow {}^4I_{11/2}$	60		2.0	60		2.0
$^4S_{3/2} \rightarrow {}^4I_{9/2}$	120		4.1	119		4.0
$^4S_{3/2} \rightarrow {}^4F_{9/2}$	1.1		0.0	1.1		0.0

(continued)

Table 9.4 (continued)

Emission band	0.25Er			1.00Er		
	A_{ED}	A_{MD}	β	A_{ED}	A_{MD}	β
τ_{cal}		0.34			0.34	
$^4F_{9/2} \rightarrow {}^4I_{15/2}$	2994		91.7	2890		91.6
$^4F_{9/2} \rightarrow {}^4I_{13/2}$	149		4.6	142		4.5
$^4F_{9/2} \rightarrow {}^4I_{11/2}$	112		3.4	112		3.6
$^4F_{9/2} \rightarrow {}^4I_{9/2}$	10		0.3	9.1		0.3
τ_{cal}		0.31			0.32	

and inhomogeneous broadening of emission spectra, among other effects. Notably, if the environment around REIs can be precisely controlled to achieve specific quantum properties, it becomes possible to tailor these materials for targeted technological applications.

The Hamiltonian REIs, see Eq. (9.1), include contributions from electrostatic $\left(\sum_{j<k}^{N} \frac{e^2}{r_{j,k}}\right)$, spin–orbit $\left(\sum_{j}^{N} \xi(r_j)(S_j \cdot L_j)\right)$, and crystal-field ($V_{CF}$) interactions. Those interactions are essential for achieving a quantitative description of the energy levels and transition probabilities of REIs. Here, the electrostatic and spin–orbit interactions are, in general, dominant in more than 80% the rest can be attributes the crystal-field interactions. Here's a short formal quantum mechanical breakdown of those contributions:

1. Trivalent REIs are multi-electron atoms, where electrons interact with each other via electrostatic repulsions which are studied using the Slater–Condon parameters, denoted by F^0, F^2, F^4, F^6.... Such parameters arise from the expansion of the Coulomb interaction $\frac{1}{r_{j,k}}\left(= \sum_{i=0}^{\infty} \frac{r_<^i}{r_>^{i+1}} P_i(\cos(\omega))\right)$, between two electrons in terms of spherical harmonics. Here $r_<^i$ is min (r_1, r_2), $r_>^{i+1}$ is max (r_1, r_2), P_i is the Legendre polynomial or order i, and ω is the angle between r_1 and r_2. The Slater–Condon F^i (direct terms) and G^i (exchange terms) parameters, for an arbitrary l^n configuration, are: $F^i = \langle ll|\frac{r_<^i}{r_>^{i+1}}|ll\rangle$, and $G^i = \langle ll'|\frac{r_<^i}{r_>^{i+1}}|l'l\rangle$ where i runs from 0 to 2 l (due to the triangle condition in angular momentum coupling) [49, 50]. Particularly, for $4f$ electrons shell, usually F2, F4, and sometimes F^6 are the most important. Here F^0 is the average Coulomb repulsion between two electrons in the same shell, and F2, F4, F^6 represent more detailed angular dependencies of the repulsion and cause the multiplets splitting of energy levels.
2. Spin–orbit coupling splits atomic energy levels that would otherwise be degenerate, thus giving rise to the so-called fine structure of atomic spectra. Such coupling is a relativistic interaction between a particle's spin and its motion in a potential. In trivalent REIs, the spin–orbit interaction plays a key role in determining energy level structure and transition probabilities. For instance, terms like ^4F split into multiple J-levels: e.g., $^4F_{3/2}$, $^4F_{5/2}$, $^4F_{7/2}$, $^4F_{9/2}$, depending on allowed $J = L + S, L + S - 1, \ldots, |L - S|$. $\xi(r)$ is the spin–orbit coupling parameter,

which depends on the radial distribution of the electron and increases with atomic number. In multi-electron atoms $\xi(r)$ is often treated as an empirical parameter. For instance, for a single electron in an orbital nl we have $\xi(r) = \frac{\hbar^2}{2m_e^2c^2}\langle R_{nl}(r)\rangle$, where $R_{nl}(r)$ is the radial wavefunction [51].

3. The crystal-field theorydescribes the electronic structure and properties of transition metal and trivalent REIs when they are placed in a crystalline or ligand environment, though with some key differences due to their unique electronic configurations. Particularly, lanthanides are shielded by outer 5s2 and $5p^6$ electrons which makes the $4f$ electrons less sensitive to the surrounding crystal field compared to transition metals where d-electrons are more exposed. In this manner, the weak crystal field produce a Stark effect on the trivalent REIs them lifts some splitting degeneracies in the J-multiplets, which is crucial for understanding optical and magnetic properties. Here, we center our attention on the optical properties of trivalent rare earth, where the largest part of the intensity studies has been limited to transitions between J-manifolds (in solutions, glasses or even in single crystals). It is important to mention that the $4f$ orbitals experience much weaker splitting (~100 cm^{-1}) compared to spin–orbit coupling (~1000 cm^{-1}). In this way, parametrization of the transition intensities between crystal-field levels (in single crystals) is also possible. Such parametrization in terms of $B_{kq}([ML^2T^{-2}])$ parameters and $A^{\lambda}_{kq}([L^{-k}])$ coefficients is difficult to solve in comparison with the parametrization of transitions between J-manifolds. The $B_{\lambda kq}$ include the radial integral r^k. The crystal-field parametrization has been described in detail by Gorller-Walrand and Binnemans (1996) [52]. The crystal-field Hamiltonian is defined as [52]: $V_{CF} = \sum_{k,q} A_{kq}\widehat{\boldsymbol{D}}^{(k)}_q = \sum_{k,q} A_{kq} \sum_j r_j^k \widehat{\boldsymbol{C}}^{(k)}_q(\theta_j, \varphi_j)$. Here $\widehat{\boldsymbol{D}}^{(k)}_q$ is a tensor of rank k with components q, r_j^k is the position of electron j (to the kth power) and $\widehat{\boldsymbol{C}}^{(k)}_q(\theta_j, \varphi_j)$ is the irreducible tensor operator of rank k containing the angular coordinates of electron j.

9.4.1 Mixing of 4f and Ligand Orbitals

In this context, the covalency model of hypersensitivity (or pseudo-quadrupole transitions) is another theoretical framework employed in trivalent REIs to explain the unusual intensity variations observed in certain electronic transitions, particularly $4f$–$4f$ transitions, in their absorption and emission spectra, which deviate from the predictions of pure electrostatic (Judd–Ofelt) theory [1, 2]. Such a model emerged from the charge-transfer transitions in lanthanide complexes, where the energies and intensities of these transitions proved highly sensitive to both ligand type and the lanthanide ion. The energy of the ligand-to-metal charge transition depends on metal orbital energy, interelectronic repulsion energy, and ligand energy. R.D. Peacock (1977) emphasizing the strong influence of the ligand on the charge-transfer energy of a single metal ion [53]. To elucidate hypersensitivity behavior, we assume ED transitions, where a perturbation is required to mix states of parity other than f^n with states of the f^n configuration. In particular, charge transfer transitions, in terms of energy and oscillator strengths, exhibit high sensitivity to ligand type and environment [54, 55].

In amorphous materials, covalent bonds form a continuous, albeit disordered, network. These bonds are crucial for maintaining the structure of the material, even in the absence of long-range order, therefore, when such materials are doped with REIs, such ions experience a wide distribution of local environments (bond lengths, angles, and ligand symmetries) [3, 17, 55]. Furthermore, these bonds involve the sharing of electron pairs between atoms, resulting in a stable three-dimensional network that gives glasses their characteristic properties. Unlike crystalline solids with long-range order, glasses exhibit short-range order where atoms are strongly bonded to their nearest neighbors, forming a continuous random network. In addition, stronger covalency implies greater overlap between the rare-earth ion's $4f$ orbitals and the ligand orbitals, including those of opposite parity-states ($4f$–$5d$ or $4f$–$4f$ with odd-order crystal-field components). Here, hypersensitivity is characterized by an increased intensity of specific transitions owing to the different components within the glass, resulting in a mixture of factors including the covalency of the ligand–metal bond and the local symmetry around the trivalent rare-earth ion. Where the intensities of $4f$–$4f$ band transitions increase in non-centrosymmetric systems by mixture, e.g., of a degree of d orbital character with the $4f$ states.

These transitions typically follow selection rules of $\Delta L \leq 2$, $\Delta J \leq 2$, and $\Delta S = 0$ [11, 14, 56], e.g., in glasses we can see the ${}^4I_{9/2} \rightarrow {}^4G_{5/2}$ transition in $Nd3^+$ with emission center ~580 nm, ${}^5D_0 \rightarrow {}^7F_{2/5}$ transition in $Eu3^+$ with emission center ~615 nm or ${}^4I_{15/2} \rightarrow {}^2H_{11/2}$ transition in $Er3^+$ with emission center ~521 nm, which are anomalously high sensitivity to the ligand environment—these are called hypersensitive transitions.

As mentioned above, the wavefunction of the ions is no longer purely $4f$, but includes contributions from ligand orbitals by opposite parity-states or by the environment around ions, e.g., ligand polarizations effects. This can be written as

$$\left|\Psi_{4f}\right\rangle = c_1|4f\rangle + c_2|L\rangle \tag{9.12}$$

where $|4f\rangle$ is the pure $4f$ orbital, $|L\rangle$ are the ligand orbitals and c_1 and c_2 are the mixing coefficients ($c_2 \ll c_1$, but non-negligible). Then, the transition moment for an ED transition between states $|\Psi_i\rangle$ and $\left|\Psi_f\right\rangle$ is

$$\langle\Psi_f|\hat{\mu}_{ED}|\Psi_i\rangle \approx c_1^2\langle 4f|\hat{\mu}_{ED}|4f\rangle + c_1c_2\langle 4f|\hat{\mu}_{ED}|L\rangle + \text{higher order terms} \tag{9.13}$$

The second term (cross term) allows weak ED transitions due to partial ligand character and depend on asymmetric ligand fields (low symmetry enhances mixing).

Now we need to describe the interaction between a metal ion ($|4f_i\rangle$) and its surrounding non-symmetric ligands $\left(\langle L_j|\right)$, that is, a similar environment within glass, where the trivalent rare earth can be represented by non-centrosymmetric systems. Furthermore, let us assume that electronic transitions arise from this system and that we can apply the covalency model of hypersensitivity. Thus, the Hamiltonian (H_{cmh}) governing the interaction (static/dynamic) is represented as

$$H_{cmh} = \sum_{i,j} \beta_{ij}\left(|4f_i\rangle\langle L_j| + |L_j\rangle\langle 4f_i|\right) \tag{9.14}$$

This term is crucial in perturbation theory to explain hypersensitivity in the covalence model. This is derived from molecular orbital (MO) theory and perturbation theory. Here β_{ij} is the covalent mixing parameter and arises from orbital overlap and energy matching between the lanthanide 4f orbitals and ligand orbitals. It is formally defined as $\beta_{ij} = \frac{\langle 4f_i|H|L_j\rangle}{E_{4f}-E_L}$, where $\langle 4f_i|H|L_j\rangle$ are the Hamiltonian matrix element (overlap integral + effective potential), E_{4f} and E_L are the energy of the 4f orbital and Energy of the ligand orbital (L_j), respectively. Thus, the 4f state gain ligand character, allowing ED transition as

$$\langle\Psi_f|\widehat{\mu}_{ED}|\Psi_i\rangle \approx \beta\langle L|\widehat{\mu}_{ED}|4f\rangle \neq 0 \tag{9.15}$$

We see then that the H_{cmh} is the ligand field Hamiltonian represents the electrostatic (i.e., Coulombic interaction between ligand point charges) and covalent effects of the ligands (charge transfer states) on the lanthanide ions [8, 57, 58]. It is important to mention that the last equation quantifies how strongly the ligand field mixes the 4f states with higher-energy states (Ψ_k), enabling formally forbidden transitions. In addition, the H_{cmh} includes also the non-electrostatic contributions (orbital overlap), which are key for explaining anomalous intensities as well as the polarizability of ligands. This model is important in low-symmetry systems (e.g., due to inherent disorder in glasses different local structural environments we found C2v and C4v symmetries) where ligand fields can induce significant 4f-ligand hybridization. In this context, techniques as X-ray scattering microscopy, EXAFS, and XANES are used to probe the local structures and symmetry of glasses [59]. These techniques help researchers understand how the local structure changes with different compositions, processing conditions, thermal treatments, and others.

In Table 9.5, we can see a comparison between the Judd–Ofelt theory and covalency model.

Diverse works in glasses have shown a correlation between the intensity of hypersensitive transitions and the covalency of the rare-earth ion's bonding with different ligands, since the degree of covalency can influence the intensity of the hypersensitive transitions. In essence, the covalency model provides a framework for understanding how the nature of the chemical bond between lanthanide ions and its surrounding ligands affects the intensity of specific electronic transitions, particularly those exhibiting hypersensitivity [60–68].

In Table 9.6, we can find a summary of diverse factors on effect on hypersensitivity.

Table 9.5 Comparison between Judd–Ofelt theory and covalency model

Feature	Judd–Ofelt theory	Covalency model
Mechanism	Odd-parity crystal field	$4f$-ligand orbital mixing
Ligand Sensitivity	Weak	Strong
Symmetry Role	Even-rank crystal fields	Anisotropic covalency
Parameters	$\Omega_2, \Omega_4, \Omega_6$	Covalency integrals β_{ij}
Correlation	The Judd–Ofelt intensity parameter Ω_2 is strongly correlated with covalency and asymmetry	Glasses with higher Ω_2 show stronger hypersensitivity transition (e.g., tellurite > borate > silicate glasses)

Table 9.6 Summary of various factors on the effect on hypersensitivity

Factor	Effect on Hypersensitivity
Ligand polarizability	↑ Covalency → ↑ Intensity
Metal–ligand bond distance	Shorter bonds → ↑ Mixing
Oxidation state	Higher oxidation (e.g., Ln^{4+} vs. Ln^{3+}) → ↑ Covalency
Symmetry	Low symmetry (e.g., C2v$C2v$) → ↑ Intensity
Non-bridging oxygen (NBO) in glasses	Increase covalency → Enhance hypersensitivity transitions

Since the intensities of hypersensitivity transitions are very sensitive to the local ligand field, they can serve as a detector of structural changes in the glass matrix.

9.4.2 Nephelauxetic Effect

Depending on the glass, it is possible to observe in the absorption or emission spectra a redshift (lower energy) at peak of the 4f–4f transition energies, this is called the nephelauxetic effect and is directly related to the covalency of the ligand–metal bond. Henrie et al. (1976) revealed a connection between hypersensitivity and the nephelauxetic effect, which was characterized not only by a redshift in 4f–4f transitions but also that increases with a greater covalency [66, 69, 70]. Bonds covalently reduce electron–electron repulsion in 4f orbitals, and a reduction in Racah parameters (B which is related to interelectronic repulsion), then a shift in energy levels and transition intensities. Some articles on Eu^{3+}, Nd^{3+}, and Sm^{3+} in glasses show slight changes in B values compared to aquo ions [47, 71]. This effect is typically attributed to an expansion of the ionic wavefunctions, suggesting a potential link between the nephelauxetic effect and the covalency as described by the covalent approach.

It should be noted that the spectroscopic features (e.g., absorption, emission, EPR from trivalent REIs) associated with these low-symmetry structures can be used to probe the local structure and dynamics of glasses.

The bonding parameter (δ) [72, 73] is defined as

$$\delta = \left(\frac{(1-\overline{\vartheta})}{\overline{\vartheta}}\right) \times 100 \tag{9.16}$$

$\overline{\vartheta} = \sum_N \vartheta/N$ and $\vartheta = E_f/E_i$ is the nephelauxetic ratio, here E_f and E_i are the energies of corresponding transitions in the complex and aquo-ion, N refers to the number of levels used to compute $\overline{\vartheta}$ values. δ can be positive or negative, indicating covalent or ionic bonding that will depend largely on the network glass. For instance, in oxide glasses (e.g., Silicate or Borate), a higher nephelauxetic effect due to stronger Ln–O covalency [63, 74, 75], in fluoride glasses (e.g., ZBLAN): here weaker effect because of more ionic Ln–F bonds [76, 77], and in chalcogenide glasses (e.g., Ge-As-Se), here a stronger covalency leads to a more pronounced effect [78].

According to P. Babu and C. K. Jayasankar, the transition energies for Dy^{3+}-doped borate glasses are relatively higher than those found for Dy^{3+}:crystalline hosts. In addition, they observed that the energy positions of Dy^{3+} ions obey the next order of magnitude: *Glasses > Aquo-ion > Crystalline hosts.*

In this way, Sutrisno et al. [66] claim that nephelauxetic effect can be used to calculate the covalency of the bonding between the rare-earth ion and the surrounding oxygens in the glass matrix.

The hypersensitivity impact on the electronic transitions of the trivalent rare-earth ions is due to influencing the mixing of electronic states and the probability of transitions, covalency plays a crucial role in determining the intensity of hypersensitive transitions. Besides, a stronger covalent bond, as indicated by a larger nephelauxetic effect, generally leads to more intense hypersensitive transitions.

9.4.3 Transition Intensity via Perturbation Theory

To determine the intensity contribution of a hypersensitivity-transition we employ the second-order perturbation theory, then:

$$I \propto \left| \sum_k \frac{\Psi_f|H_{cmh}|\Psi_k \Psi_k|\hat{\mu}_{ED}|\Psi_i}{E_i - E_k} \right|^2 + \textit{higher order terms contributions} \tag{9.17}$$

H_{cmh} is the ligand-asymmetric fields Hamiltonian that describe how the ligand environment modifies the trivalent rare-earth ion's electronic structure. Besides, its role is to mediate intensity borrowing from higher-energy states, odd mixture, making Laporte-forbidden $4f$–$4f$ transitions partially allowed.

Actually, understanding the structure–property relationships in low-symmetry systems like glasses is decisive to design and optimize materials with specific optical, electronic, and magnetic properties for various technological applications. For instance, in laser materials: tuning hypersensitivity transitions via ligand design (e.g., $Nd3^+$ in YAG vs. glasses); as optical amplifiers: $Er3^+$-doped fluoride glasses (low-phonon energy + hypersensitive transitions) for telecom; in sensors: hypersensitive transitions as probes for local symmetry (e.g., bio-imaging with RE nanoparticles); and as quantum materials: an engineering $4f$–$5d$ mixing for spin–orbit coupling control, coherence and other properties.

9.5 Role of Symmetry—Crystal-Field Theory in Trivalent Rare-Earth Ions

In a free lanthanide ion, the $4f$ orbitals are degenerate, all the $4f^n$ electrons with the same energy levels. Nevertheless, the local symmetry of the ligand field (which is non-uniform in glass) lifts this degeneracy, leading to Stark splitting.

Glasses doped with trivalent REIs play an important role in modern technology industries as well as the quantum technological development. Here, the performance, efficiency, and region of the absorption or emission of light are tuned via the choice of the lanthanide ions, the transitions within different energy levels of the luminescent ions or band-to-band transitions of the host compound or the other glass-network components. Therefore, it is important to know and predict the luminescence property of the doped glasses before spectrometer measurement. In this context, crystal-field theory describes the Stark effect of outer-shell $4f^n$ electrons embedded in the electrostatic potential field originating from the neighboring ligands.

To arrive formally at an expression for CF, one can think of a trivalent rare-earth ion in a sphere about the origin with charges surrounding [79], e.g., O^{2-} in oxide glasses or F^- in fluoride glasses [52]. Such charges create a crystal-field potential that can split the degenerate $4f^n$ energy levels. Then the potential due to the rest of the lattice, at spherical coordinates, is

$$V_{CF}(r,\theta,\phi) = \sum_{i=0}^{\infty}\sum_{m=-l}^{l} A_l^m r^l Y_l^m(\theta,\phi) \tag{9.18}$$

A_l^m are the CF parameters, units: energy, e.g., cm¯1, and depends on ligand positions and charges, Y_l^m are the l-rank spherical harmonics. The A_l^m parameters are defined as

$$A_l^m = \sum_i Z_i e^2 \left(\frac{Y_l^m(\theta_i', \phi_i')}{R_i^{n+1}} \right) \tag{9.19}$$

A_l^m is over all the i charges Z_i is the net charge of the Lth ligand, R_i is the distance from the central lanthanide ion with each surrounding. For $4f$ electrons ($l = 3$), the CF expansion is usually truncated at $l = 6$. Besides, A_l^m is nonzero only when $l =$ 0, 2, and 4, and the values m are limited by the point symmetry. It is important to mention that the symmetry of the site determines the CF terms that are needed and the form of the radial in Eq. (9.19) is significant for the magnitude. Glasses lack good symmetry, therefore all allowed A_l^m terms contribute. Subsequently glass, the CF splitting is inhomogeneously broadened, leading to multiple overlapping sub-levels, both of absorption/emission lines. For instance, in silica glasses it is well known that the broadband is around 35 nm [80], but in oxide-tellurite or fluorotellurite glasses the broadband is more than 60 nm [81].

For instance, in fluoride glasses (ZBLAN), the non-radiative relaxation we can see that phonon-assisted transitions depend on CF-induced energy gaps, e.g., Pr^{3+}, Dy^{3+}. On the other hand, in chalcogenide Glasses a strong CF effect is observed because of the high polarizability of S/Se ligands.

9.5.1 *Stevens Operator Formalism*

In glasses, the symmetry around the lanthanide ions can often be too complicated, so we must try to replace $Y_l^m(\theta, \phi)$ with polynomials in angular momentum operators $\left(J_x, J_y, J_z\right)$ on a constant J manifold. In this case, the Stevens operator formalism, introduced by K.W.H. Stevens in 1952, is the most suitable method. This formalism is particularly useful for describing CF Hamiltonians and for analyzing spectroscopic data in materials science and condensed matter physics, such as EPR, ESR, and Mössbauer spectroscopy. Here, the CF Hamiltonian is often expressed as O_l^m, which are angular momentum operators acting on the $|J, m_J\rangle$ states of the trivalent rare-earth ion [82]:

$$V_{CF} = \sum_{l=2,4,6} \sum_{m=-l}^{l} A_l^m O_l^m \left(J_x, J_y, J_z\right) \tag{9.20}$$

Here O_l^m is the Stevens operators. As mentioned above, in glasses the CF lacks long-range order, thus all Aml terms are allowed, and the CF Hamiltonian take a statistically distributed, leading to inhomogeneous broadening of spectral lines, this can be expressed also with the Stevens operators. In addition, the energy levels are found by diagonalizing:

$$V_{CF}|J, m_J\rangle = \mathrm{E}|J, m_J\rangle \tag{9.21}$$

The degeneracy of the states is partially or fully lifted. Next, we will see the list of Stevens operators, for $l = 2$, 4, and 6 since these are the terms that contribute significantly to the Hamiltonian [83], see Table 9.7.

Table 9.7 List of Stevens operators for second, fourth, and sixth orders

m	$l = 2 \rightarrow O_2^m$	$l = 4 \rightarrow O_4^m$	$l = 6 \rightarrow O_6^m$
0	$O_2^0 = 3J_z^2 - J(J+1)$	$O_4^0 = 35J_z^4 - [30J(J+1) - 25]J_z^2 + 3J^2(J+1)^2 - 6J(J+1)$	$O_6^0 = 231J_z^6 - (315J(J+1) - 735)J_z^4 + \left(105J^2(J+1)^2 - 525J(J+1) + 294\right)J_z^2 - 5J^3(J+1)^3 + 40J^2(J+1)^2 - 60J(J+1)$
± 1	$O_2^{\pm 1} = \frac{1}{2}(J_\pm J_z + J_z J_\pm)$	$O_4^{\pm 1} = \frac{1}{4}\left(7J_z^3 - 3J(J+1)J_z \mp J_z\right)J_\pm + \ldots$	$O_6^{\pm 1} = \frac{1}{4}\left(33J_z^5 - [30J(J+1) - 15]J_z^3 + \left[5J^2(J+1)^2 - 10J(J+1) + 12\right]J_z\right)J_\pm \ldots$
± 2	$O_2^{\pm 2} = j_\pm^2$	$O_4^{\pm 2} = \frac{1}{2}\left(7J_z^2 - J(J+1) - 5\right)J_\pm^2 + \ldots$	$O_6^{\pm 2} = \frac{1}{4}\left(33J_z^4 - [18J(J+1) - 123]J_z^2 + J^2(J+1)^2 - 2J(J+1) + 102\right)J_\pm^2 + \ldots$
± 3		$O_4^{\pm 3} = \frac{1}{4}\left(J_z J_\pm^3 + J_\pm^3 J_z\right)$	$O_6^{\pm 3}$ and $O_6^{\pm 4}$ are higher m terms involved $J_\pm^3$ and $J_\pm^4$, etc.
± 4		$O_4^{\pm 4} = J_\pm^4$	

The physical meaning of Stevens operators:

O_2^0 represents the quadrupolar interaction (splitting due to axial distortion)

O_4^0, O_6^0 represent the higher-order anisotropy terms (important for magnetic properties)

$O_l^{\pm m}$ **is** the mixing of m_j states (affects optical transitions)

For example, if the RE^{3+} site has approximate C_{2v} symmetry (common in glasses), the CF Hamiltonian simplifies to

$$V_{CF} = A_2^0 O_2^0 + A_2^2 O_2^2 + A_4^0 O_4^0 + A_4^2 O_4^2 + A_4^4 O_4^4 + \dots \tag{9.22}$$

A_2^0 and A_2^2 are the axial and rhombic distortion, respectively. Besides the higher-order terms A_4^m and A_6^m refine the Stark splitting. For more information and applications about CF look for [78, 84–94].

9.6 Dynamic Coupling Mechanism

Another important quantum mechanism to consider in the spectroscopic analysis of trivalent rare-earth ions is dynamic coupling mechanisms. Those interaction processes are vital not only for understanding them, since it introduces a way for these forbidden transitions to occur, but also for optimizing their luminescence and other optical properties, especially in glass materials for technological applications. These mechanisms, which can include energy transfer processes between ions and interactions with the host material as well as other rare earth inside the glass matrix or plasmonic nanostructures resulting in quantum-plasmonic interactions.

This dynamic coupling was firstly proposed by Jörgensen and Judd [60] to explain the unusual intensity variation of certain $4f$–$4f$ transitions designated hypersensitive transitions. We must be harassed here, in contrast with a common procedure found in the literature in the case of glasses doped with trivalent rare-earth ions, the dynamic coupling mechanism cannot, in any circumstances, be neglected.

In a light–matter interaction, the dynamic coupling is manifested as the modulation of the crystal field by lattice vibrations (phonons, polarized ions, or oscillation charges). This leads to:

- Broadening of spectral lines (homogeneous/inhomogeneous), including changes lineshape.
- Non-radiative relaxation (multiphonon decay, i.e., non-radiative decay).
- Sideband features in absorption/emission spectra.
- Changes in the lifetime and thus quantum yield.

9.6.1 *Hamiltonian Including Dynamic Coupling*

In this manner, the total Hamiltonian for a trivalent rare-earth ion, inside a glass with a vibrating lattice, a distribution of non-centrosymmetric systems, ligand charges, and under a light–matter interaction can be written as

$$H = H_{electronic} + V_{CF} + H_{cmh} + H_{phonon} + H_{electron-phonon} + H_{\text{ligand-polarization}} \tag{9.23}$$

where:

$H_{electronic} = -\frac{\hbar^2}{2m}\sum_j^N \nabla_j^2 - \sum_j^N \frac{Ze^2}{r_j} + \sum_{j<k}^N \frac{e^2}{r_{j,k}} + \sum_j^N \xi(r_j)(S_j \cdot L_j)$, considering the Judd–Ofelt theory.

$V_{CF} = \sum_{l=2,4,6} \sum_{m=-l}^{l} (A_l^m + \Delta A_l^m) O_l^m (J_x, J_y, J_z)$, is the static and dynamic CF potential, ΔA_l^m represents phonon-induced fluctuations.

H_{cmh} is defined in Eq. (9.14).

$H_{phonon} = \sum_q \hbar\omega_q \left(b_q^+ b_q + \frac{1}{2}\right)$ is the contribution grom the phonon modes, i.e., quantified vibrational modes in a crystal lattice, representing the collective oscillations of atoms. Here b_q^+ and b_q are the phonon annihilation and creation operators, respectively.

$H_{electron-phonon} = \sum_{l,m,q} V_{l,m}(q) O_l^m \left(b_q^+ + b_q\right)$, dynamic electron–phonon coupling. Where $V_{l,m}$ is the coupling strength for phonon mode q, which is related with Huang–Rhys factor S. Such S is a dimensionless parameter and represents the average number of phonons involved in an electronic transition. If $S = 0$, no phonon coupling, thus, purely electronic transition. But if $S \gg 1$ we have strong coupling, where broad, phonon-assisted transitions dominate [95, 96]. Nevertheless, trivalent lanthanides ions typically exhibit small S values ($S \ll 1$) because $4f$ electrons are shielded by $5s^2$and $5p^6$, reducing electron–phonon coupling. Then, optical transitions are mostly sharp (weak phonon sidebands). For instance, S for Er^{3+}-doped silicate glass with a $4I13/2 \leftrightarrow 4I15/2$ (at 1.5 μm), typical S is ~0.1–0.3, hence a weak coupling with phonon energy ~300 cm^{-1} (Si–O vibrations) [97].

$H_{\text{ligand-polarization}} = \sum_p \alpha_p \widehat{Q}_p \cdot \boldsymbol{E}(t)$, is the dynamic coupling mechanism which involves ligand polarization induced by the oscillating electric field of light $\boldsymbol{E}(t)$. Here α_p is the polarizability tensor, which describes how an atom, ion, or molecule distorts its electron cloud in response to an external electric field. In isotropic such materials, like glasses, α_{ij} reduces to a scalar $\alpha_{ij} = \alpha\delta_{ij}$. $\widehat{Q}_p$ is the ligand multipole operator, e.g., dipole and/or quadrupole order. It is important to mention that this polarization modifies the effective crystal-field potential experienced by the $4f^n$ electrons. Moreover, the oscillating electric field of the lanthanide's ion can induce dipoles in nearby ligands, especially in polarizable glasses like fluorides of chalcogenides [96]. It should be noted that the polarization can screen or enhance certain CF components, e.g., in fluoride glasses (ZBLAN), F^- is less polarizable than O^{2-}, leading to narrower linewidths.

Rivera et al. [98–102] argument that plasmonic nanostructures with trivalent rare-earth ions nearby can interact via a quantum–plasmonic interaction, in this case via a dynamic coupling because the plasmonic nanostructures represents charge oscillations, both localized or propagating, enable of modified the electronic transitions, lifetime and other effects, see, for example, [99, 100, 102].

When dealing with energy transfer processes between REIs, we have dipoles oscillating at different frequencies, with resonant or non-resonant energy levels. It must be considered that in the dipole–dipole or dipole–quadrupole expressions for transfer rates, the separation distances between them depend on the concentration of each of them, the pumping power of the laser, and no less importantly, the type of glass matrix used. Precise control of this mechanism makes it possible, for example, to obtain white light, thermometric sensors, and increased bandwidth, among others [4, 5, 31, 103–106].

Several interesting aspects may be discussed about the forced ED and dynamic coupling mechanisms in glasses, please see the next references to see more examples and applications on dynamic coupling [58].

The eigenvalues, energies of each of the interactions described in Eq. (9.23), are determined phenomenologically from experimental intensities, through emission, absorption, FTIR or Raman spectra, among others, and many times such interactions are coupled simultaneously, competitive processes, and cannot be distinguished from each other. This is one of the reasons why the theoretical calculations of the contributions of individual terms are motivated. However, such theoretical calculations in glassy materials prove to be a bigger challenge, because of the wide variety of site symmetries occupied by the lanthanide ions.

References

1. B.R. Judd, Phys. Rev. **127**, 750 (1962)
2. G.S. Ofelt, J. Chem. Phys. **37**, 511 (1962)
3. L.S. Archana, D.N. Rajendran, J. Cyriac, J. Lumin. **243**, 118679 (2022)
4. B. Bondzior, C. Nguyen, T.H. Quan Vu, D. Pugliese, P.J. Dereń, L. Petit, J. Lumin. **252**, 119386 (2022)
5. G. Lozano, O.B. Silva, W. Faria, A.S.S. de Camargo, R.R. Gonçalves, D. Manzani, R. Bruna, V.A.G. Rivera, E. Marega, J. Lumin. **228**, 117538 (2020)
6. C. Yu, X. Zhang, X. Li, J. Zhang, S. Xu, X. Zhang, Y. Zhang, X. Wang, L. Wang, G. Sui, B. Chen, Mater. Res. Bull. **120**, 110590 (2019)
7. F. Auzel, S. Hubert, P. Delamoye, J. Lumin. **26**, 251 (1982)
8. R.T. Moura, A.N. Carneiro Neto, R.L. Longo, O.L. Malta, J. Lumin. **170**, 420 (2016)
9. A. Ćirić, Ł Marciniak, M.D. Dramićanin, Sci. Rep. **12**, 563 (2022)
10. A. Kornienko, P. Loiko, E. Dunina, L. Fomicheva, A. Kornienko, Opt. Mater. **148**, 114808 (2024)
11. V.A.G. Rivera, L.A. de O. Nunes, Technol. Adv. Tellurite Glas. Prop. Process. Appl. 41 (2017)
12. G. Lozano C., J. Chacaliaza-Ricaldi, J.L. Clabel H., Y. Messaddeq, V.A.G. Rivera, E. Marega, J. Lumin. **260**, 119853 (2023)
13. Z. Pan, P. Loiko, S. Slimi, H. Yuan, Y. Wang, Y. Zhao, P. Camy, E. Dunina, A. Kornienko, L. Fomicheva, L. Wang, W. Chen, U. Griebner, V. Petrov, R.M. Solé, F. Díaz, M. Aguiló, X. Mateos, J. Lumin. **246**, 118828 (2022)
14. W.T. Carnall, P.R. Fields, B.G. Wybourne, J. Chem. Phys. **42**, 3797 (1965)
15. M. Shoaib, I. Khan, G. Rooh, S.M. Wabaidur, M.A. Islam, N. Chanthima, S. Kothan, I. Ullah, A. Ahad, J. Kaewkhao, J. Lumin. **247**, 118884 (2022)
16. M.G. Brik, A.M. Srivastava (eds.), in *Lumin. Mater.* (De Gruyter, 2023), pp. 19–62
17. D. Piatkowski, S. Mackowski, Opt. Mater. **34**, 2055 (2012)

18. S.Y. Marzouk, M.A. Azooz, H.A.E. Batal, J. Mol. Struct. **1243**, 130925 (2021)
19. M.A. Farag, A.A. Elbakey, M.K. El-Mansy, T.Y. Elrasasi, Opt. Quantum Electron. **55**, 358 (2023)
20. O.B. Silva, V.A.G. Rivera, Y. Ledemi, Y. Messaddeq, E. Marega, J. Lumin. **232**, 117808 (2021)
21. M.M. EL-Hady, H.Y. Morshidy, M.A. Hassan, J. Lumin. **263**, 119972 (2023)
22. Y. Pepe, M. Erdem, A. Sennaroglu, G. Eryurek, J. Non-Cryst, Solids **522**, 119501 (2019)
23. R.A. Tafida, M.K. Halimah, M.N. Azlan, S.B. Adamu, I.I. Lakin, Optik **262**, 169370 (2022)
24. S.V. Ushakov, Q.-J. Hong, A.I. Pavlik, A. van de Walle, A. Navrotsky, Chem. Mater. **36**, 4868 (2024)
25. X. Shen, Y. Zhang, L. Xia, J. Li, G. Yang, Y. Zhou, Opt. Laser Technol. **129**, 106264 (2020)
26. J. Liu, X. Huang, H. Pan, X. Zhang, X. Fang, W. Li, H. Zhang, A. Huang, Z. Xiao, J. Alloys Compd. **866**, 158568 (2021)
27. F. Aouaini, A. Maaoui, N.B.H. Mohamed, M.M. Alanazi, L.A. El Maati, J. Alloys Compd. **894**, 162506 (2022)
28. Y. Wu, C. Niu, L. Wang, M. Yang, S. Zhang, J. Lumin. **266**, 120323 (2024)
29. M. Łukaszewicz, R. Lisiecki, B. Klimesz, W. Ryba-Romanowski, Mater. Res. Bull. **132**, 110996 (2020)
30. D. Zhou, Q. Jiang, C. Chen, H. Zeng, D. Chen, J. Non-Cryst, Solids **583**, 121468 (2022)
31. B. Klimesz, R. Lisiecki, W. Ryba-Romanowski, J. Alloys Compd. **823**, 153753 (2020)
32. H.S. Albaqawi, S.M. AL-shomar, I.M. Danmallam, I. Bulus, J. Alloys Compd. **965**, 171101 (2023)
33. M. Azam, D.K. Mohanty, V.K. Rai, K. Singh, J. Lumin. **239**, 118319 (2021)
34. T. Wu, J. Hu, K. Han, C. Zhang, M. Li, R. Pang, Y. Hu, D. Zhou, Ceram. Int. **49**, 11718 (2023)
35. Y. Yamsuk, P. Yasaka, J. Keawkhao, N. Sangwaranatee, J. Phys. Conf. Ser. **1428**, 012028 (2020)
36. J. Rare Earths **39**, 33 (2021)
37. Mater. Res. Bull. **147**, 111651 (2022)
38. A.A. Kornienko, A.A. Kaminskii, E.B. Dunina, Phys. Status Solidi B **157**, 267 (1990)
39. J. Alloys Compd. **845**, 156028 (2020)
40. M. Kumar, R.P. Vijayalakshmi, Y.C. Ratnakaram, J. Mol. Struct. **1265**, 133333 (2022)
41. M.K. Poojha, K.A. Naseer, P. Matheswaran, K. Marimuthu, E. El Shiekh, Opt. Mater. **143**, 114289 (2023)
42. L.H.C. Andrade, S.M. Lima, R.A. Silva, M.R. Devecchi, J.C.S. Moraes, D. de Ligny, L.A.O. Nunes, J. Lumin. **246**, 118812 (2022)
43. Chem. Phys. **528**, 110513 (2020)
44. P. Evangelin Teresa, K. Marimuthu, Opt. Mater. **122**, 111704 (2021)
45. P. Suthanthirakumar, S. Arunkumar, K. Marimuthu, J. Alloys Compd. **760**, 42 (2018)
46. S.B. Adamu, M.K. Halimah, K.T. Chan, F.D. Muhammad, S.N. Nazrin, R.A. Tafida, J. Lumin. **250**, 119099 (2022)
47. A. Kaminskii, *Crystalline Lasers: Physical Processes and Operating Schemes* (CRC Press, 2020)
48. F. Auzel, in *Encycl. Mater. Sci. Technol.*, edited by K.H.J. Buschow, R.W. Cahn, M.C. Flemings, B. Ilschner, E.J. Kramer, S. Mahajan, P. Veyssière (Elsevier, Oxford, 2001), pp. 9543–9551
49. J.C. Slater, *Quantum Theory of Atomic Structure.* (New York : McGraw-Hill, 1960)
50. R.D. Cowan, *The Theory of Atomic Structure and Spectra* (University of California Press, 1981)
51. J.J. Sakurai and J. Napolitano, *Modern Quantum Mechanics* (Cambridge University Press, 2017)
52. C. Görller-Walrand, K. Binnemans, in *Handb. Phys. Chem. Rare Earths* (Elsevier, 1996), pp. 121–283
53. R.D. Peacock, Mol. Phys. **33**, 1239 (1977)

54. P.A. Ryabochkina, S.A. Antoshkina, E.V. Bolshakova, M.A. Ivanov, V.V. Kochurihin, A.V. Malov, S.N. Ushakov, N.V. Shchuchkina, K.N. Nishchev, J. Lumin. **132**, 1900 (2012)
55. S. Tanabe, Comptes Rendus Chim. **5**, 815 (2002)
56. A. Balakrishna, D. Rajesh, Y.C. Ratnakaram, J. Lumin. **132**, 2984 (2012)
57. M.F. Reid, in *Handb. Phys. Chem. Rare Earths* (Elsevier, 2016), pp. 47–64
58. O.L. Malta, L.D. Carlos, Quím. Nova **26**, 889 (2003)
59. A.C.Y. Liu, E.D. Bøjesen, R.F. Tabor, S.T. Mudie, A. Zaccone, P. Harrowell, T.C. Petersen, Sci. Adv. **8**, eabn0681 (2022)
60. Chr. K. Jørgensen, B.R. Judd, Mol. Phys. **8**, 281 (1964)
61. R.D. Peacock, in *Rare Earths*, edited by E. Nieboer, C.K. Jørgensen, R.D. Peacock, R. Reisfeld (Springer, Berlin, Heidelberg, 1975), pp. 83–122
62. R. Reisfeld, L. Boehm, J. Non-Cryst, Solids **17**, 209 (1975)
63. K. Gatterer, G. Pucker, H.P. Fritzer, S. Arafa, J. Non-Cryst, Solids **176**, 237 (1994)
64. V.K. Tikhomirov, S.A. Tikhomirova, J. Non-Cryst, Solids **274**, 50 (2000)
65. K. Iftikhar, Inorganica Chim. Acta **129**, 261 (1987)
66. M.S. Sutrisno, N.N. Yusof, M.N. Azlan, M.H.M. Zaid, S.M. Iskandar, R. Hisam, Results Chem. **6**, 101085 (2023)
67. J. Rajagukguk, M. Djamal, R. Hidayat, S. Suprijadi, A. Amminudin, Y. Ruangtaweep, J. Kaewkhao, J. Pure Appl. Chem. Res. **5**, 148 (2016)
68. F. Auzel, Chem. Rev. **104**, 139 (2004)
69. D.E. Henrie, R.L. Fellows, G.R. Choppin, Coord Chem. Rev. Neth. **18**, (1976)
70. M.F. Reid, in *Handb. Phys. Chem. Rare Earths*, edited by J.-C. G. Bünzli and V. K. Pecharsky (Elsevier, 2016), pp. 47–6.
71. R. Balda, J. Fernández, M. Sanz, A. de Pablos, J.M. Fdez-Navarro, J. Mugnier, Phys. Rev. B **61**, 3384 (2000)
72. W.S. Koski, Science **139**, 747 (1963)
73. P. Babu, C.K. Jayasankar, Opt. Mater. **15**, 65 (2000)
74. Chr. K. Jørgensen, in *Prog. Inorg. Chem.* (John Wiley & Sons, Ltd, 1962), pp. 73–124
75. S.O. Adeleye, A.A. Adeleke, P. Nzerem, A.I. Olosho, E.N. Anosike-Francis, T.S. Ogedengbe, P.P. Ikubanni, R.A. Saleh, J.A. Okolie, Trends Sci. **21**, 8759 (2024)
76. J. Lucas, J. Common Met. **112**, 27 (1985)
77. G. Zhang, M.J. Poulain, J. Alloys Compd. **275–277**, 15 (1998)
78. J. Heo, J. Non-Cryst, Solids **326–327**, 410 (2003)
79. G. Burns, J. Chem. Phys. **42**, 377 (1965)
80. C.E. Chryssou, F. Di Pasquale, C.W. Pitt, IEEE J. Sel. Top. Quantum Electron. **6**, 114 (2000)
81. V.A.G. Rivera, I.C. Pinto, R. Falci, V. Fuertes, Y. Messaddeq, J. Alloys Compd. **976**, 173182 (2024)
82. P. Hoffmann, J. Phys. Math. Gen. **24**, 35 (1991)
83. O. Duros, A. Juhin, H. Elnaggar, G.S. Chiuzbăian, C. Brouder, J. Phys. Math. Theor. **58**, 025207 (2024)
84. D. Sarkar, R. Segnan, E.K. Cornell, E. Callen, R. Harris, M. Plischke, M.J. Zuckerman, Phys. Rev. Lett. **32**, 542 (1974)
85. R. Reisfeld, C.K. Jørgensen, in *Lasers Excit. States Rare Earths*, edited by R. Reisfeld and C.K. Jørgensen (Springer, Berlin, Heidelberg, 1977), pp. 64–122
86. J. Hormadaly, R. Reisfeld, J. Non-Cryst, Solids **30**, 337 (1979)
87. J.K. Lang, Y. Baer, P.A. Cox, Phys. Rev. Lett. **42**, 74 (1979)
88. Y.D. Huang, M. Mortier, F. Auzel, Opt. Mater. **15**, 243 (2001)
89. H. Inoue, K. Soga, A. Makishima, J. Non-Cryst, Solids **298**, 270 (2002)
90. K. Latkin, Opt. Commun. **542**, 129564 (2023)
91. M. Diviš, V. Nekvasil, J. Kuriplach, Phys. C Supercond. **301**, 23 (1998)
92. Z. Luo, Y. Huang, in *Phys. Solid-State Laser Mater.*, edited by Z. Luo and Y. Huang (Springer, Singapore, 2020), pp. 385–413
93. W.C. Wang, B. Zhou, S.H. Xu, Z.M. Yang, Q.Y. Zhang, Prog. Mater. Sci. **101**, 90 (2019)
94. M. Kemere, J. Sperga, U. Rogulis, G. Krieke, J. Grube, J. Lumin. **181**, 25 (2017)

95. A. Meijerink, G. Blasse, J. Sytsma, C. Donega, A. Ellens, Acta Phys. Pol. A **90**, 109 (1996)
96. J. Common Met. **126**, 187 (1986)
97. R.M. Nor, S.N.M. Halim, M.F.M. Taib, M.K. Abd-Rahman, Solid State Phenom. **268**, 92 (2017)
98. V.A.G. Rivera, Y. Ledemi, M.A. Pereira-da-Silva, Y. Messaddeq, E. Marega Jr., Sci. Rep. **6**, 18464 (2016)
99. V.A.G. Rivera, F.A. Ferri, M.H. Gehlen, L.A.O. Nunes, E. Marega, Opt. Eng. **57**, 1 (2018)
100. G. Lozano C., O.B. Silva, F.A. Ferri, V.A.G. Rivera, E. Marega, Sci. Rep. **12**, 5015 (2022)
101. V.A.G. Rivera, M. El-Amraoui, Y. Ledemi, Y. Messaddeq, E. Marega, J. Lumin. **145**, 787 (2014)
102. V.A.G. Rivera, O.B. Silva, Y. Ledemi, Y. Messaddeq, E. Marega, *Collective Plasmon-Modes in Gain Media: Quantum Emitters and Plasmonic Nanostructures* (Springer International Publishing, Cham, 2015)
103. G.L.C., V.A.G. Rivera, Y. Messaddeq, E.M.Jr., J. Alloys Compd. **948**, 169777 (2023)
104. V.A.G. Rivera, F.A. Ferri, L.A.O. Nunes, E. Marega, Opt. Mater. **67**, 25 (2017)
105. D.H. Spadoti, I.C. Pinto, V.A.G. Rivera, Y. Messaddeq, Opt. Express **33**, 10069 (2025)
106. I.C. Pinto, J.-L. Delarosbil, D.H. Spadoti, V.A.G. Rivera, Y. Messaddeq, Opt. Express **33**, 15451 (2025)

Appendix

Reduced Matrix Elements for Some Rare-Earth Ions

Reduced matrix elements $\left|\langle S, L, J|U^{\lambda}|S', L', J'\rangle\right|^2$ for Pr^{3+} (aquo), being $^{2S+1}L_J$ = $^3H_4$ the ground state. The wavelength is given in nm [1].

Upper state	λ = 2	λ = 4	λ = 6
$^3H_5$	0.0001	0.0331	0.1396
$^3H_6$	0.5081	0.4047	0.1189
$^3F_2$	0.0655	0.3474	0.6996
$^3F_3$	0.0174	0.0526	0.4958
$^3F_4$	0.0019	0.0044	0.0118
$^1G_4$	0	0.1719	0
$^1D_2$	0	0.1708	0
$^3P_0$	0.0087	0.0482	0.0220
$^3P_1$	0	0.0333	0.1247
$^3P_2$	0	0.0072	0

Reduced matrix elements $\left|\langle S, L, J|U^{\lambda}|S', L', J'\rangle\right|^2$ for Nd^{3+} (aquo), being $^{2S+1}L_J$ = $^4I_{9/2}$ the ground state. The wavelength is given in nm [1].

Upper state	λ = 2	λ = 4	λ = 6
$^4I_{11/2}$	0.0194	0.1073	1.1652
$^4I_{13/2}$	0.0001	0.0136	0.4557
$^4I_{15/2}$	0	0.0001	0.0453
$^4F_{3/2}$	0	0.2299	0.0547
$^4F_{5/2}$	0.0009	0.2361	0.3973

(continued)

G. Lozano C. et al., *A Brief Guide to Optical Spectroscopy of Trivalent Rare Earth Ions in Glasses*, Springer Series in Optical Sciences 247,
https://doi.org/10.1007/978-3-032-18501-3

(continued)

Upper state	λ = 2	λ = 4	λ = 6
$^2H_{9/2}$	0.0092	0.0076	0.1196
$^4F_{7/2}$	0	0.0027	0.2354
$^4S_{3/2}$	0.0010	0.0420	0.4249
$^4F_{9/2}$	0.0010	0.0096	0.0381
$^2H_{11/2}$	0.0001	0.0026	0.0104
$^4G_{5/2}$	0.8968	0.4091	0.0355
$^2G_{7/2}$	0.0755	0.1848	0.0316
$^2K_{13/2}$	0.0068	0.0002	0.0313
$^4G_{7/2}$	0.0551	0.1569	0.0547
$^4G_{9/2}$	0.0046	0.0611	0.0407
$^2K_{15/2}$	0	0.0053	0.0144
$^2G_{9/2}$	0.0010	0.0144	0.0136
$^2D_{3/2}$	0	0.0181	0.0001
$^4G_{11/2}$	0	0.0053	0.0080
$^2P_{1/2}$	0	0.0381	0
$^2D_{5/2}$	0	0.0001	0.0024
$^2P_{3/2}$	0	0.0030	0.0010
$^4D_{3/2}$	0	0.1954	0.0168
$^4D_{5/2}$	0.0001	0.0573	0.0280
$^2I_{11/2}$	0.0049	0.0148	0.0034

Reduced matrix elements $\left|\langle S, L, J\,|U^{\lambda}|S', L', J'\rangle\right|^2$ for Sm^{3+} (aquo), being $^{2S+1}L_J = {}^6H_{5/2}$ the ground state. The wavelength is given in nm [2].

Upper state	λ = 2	λ = 4	λ = 6
$^6H_{7/2}$	0.2062	0.1962	0.0952
$^6H_{9/2}$	0.0256	0.1395	0.3267
$^6H_{11/2}$	0	0.0240	0.2649
$^6H_{13/2}$	0	0.0007	0.0659
$^6F_{1/2}$	0.1939	0	0
$^6F_{3/2}$	0	0	0.0043
$^6H_{15/2}$	0.1444	0.1364	0
$^6F_{5/2}$	0.0332	0.2840	0
$^6F_{7/2}$	0.0020	0.1429	0.4301
$^6F_{9/2}$	0	0.0206	0.3413
$^6F_{11/2}$	0	0.0006	0.0515
$^4G_{5/2}$	0.0002	0.0007	0

(continued)

(continued)

Upper state	$\lambda = 2$	$\lambda = 4$	$\lambda = 6$
$^4F_{3/2}$	0.0003	0	0
$^4G_{7/2}$	0.0004	0.0018	0.0025
$^4I_{11/2}$	0.0022	0.0005	0.0014
$^4I_{13/2}$	0	0	0.0307
$^4I_{15/2}$	0	0	0.0108
$^6P_{5/2}$	0	0.0030	0.0228
$^6P_{3/2}$	0.0004	0.0002	0

Reduced matrix elements $|\langle S, L, J|U^\lambda|S', L', J'\rangle|^2$ for Eu^{3+} (aquo), being $^{2S+1}L_J$ = $^7F_0$ and $^7F_1$ the lowest states. The wavelength is given in nm [3]. The $^7F_1$ manifold is located about 350 cm^{-1} above the ground state.

Upper state	Transitions from $^7F_0$			Transitions from $^7F_0$		
	$\lambda = 2$	$\lambda = 4$	$\lambda = 6$	$\lambda = 2$	$\lambda = 4$	$\lambda = 6$
$^7F_1$	0	0	0			
$^7F_2$	0.1375	0	0	0.0518	0	0
$^7F_3$	0	0	0	0.2093	0.1281	0
$^7F_4$	0	0.1402	0	0	0.1741	0
$^7F_5$	0	0	0	0	0.1193	0.0544
$^7F_6$	0	0	0.1450	0	0	0.3773
$^5D_0$	0	0	0	0	0	0
$^5D_1$	0	0	0	0.0026	0	0
$^5D_2$	0.0008	0	0	0.0001	0	0
$^5D_3$	0	0	0	0.0004	0.0012	0
$^5L_6$	0	0	0.0155	0	0	0.0090
$^5G_2$	0.0006	0	0	0.0006	0	0
$^5G_3$	0	0	0	0	0	0.0183
$^5G_4$	0	0	0	0.0003	0.0014	0
$^5G_6$	0	0.0007	0	0	0.0002	0
$^5L_8$	0	0	0	0	0.0005	0.0100
$^5D_4$	0	0	0.0038	0	0	0.0051

Reduced matrix elements $|\langle S, L, J|U^\lambda|S', L', J'\rangle|^2$ for Tb^{3+} (aquo), being $^{2S+1}L_J$ = $^7F_0$ the ground state. The wavelength is given in nm [4].

Upper state	$\lambda = 2$	$\lambda = 4$	$\lambda = 6$
$^7F_5$	0.5376	0.6418	0.1175
$^7F_4$	0.0889	0.5159	0.2654

(continued)

(continued)

Upper state	$\lambda = 2$	$\lambda = 4$	$\lambda = 6$
$^7F_3$	0	0.2324	0.4126
$^7F_2$	0	0.0482	0.4695
$^7F_1$	0	0	0.3763
$^7F_0$	0	0	0.1442
$^5D_4$	0.0010	0.0008	0.0013
$^5D_3$	0	0.0002	0.0014
$^5G_6$	0.0017	0.0045	0.0118
$^5L_{10}$	0	0.0004	0.0592
$^5G_5$	0.0012	0.0018	0.0135
$^5D_2$	0	0	0.0008

Reduced matrix elements $\left|\langle S, L, J|U^\lambda|S', L', J'\rangle\right|^2$ for Dy^{3+} (aquo), being $^{2S+1}L_J = {}^6H_{15/2}$ the ground state. The wavelength is given in nm [2].

Upper state	$\lambda = 2$	$\lambda = 4$	$\lambda = 6$
$^6H_{13/2}$	0.2457	0.4139	0.6824
$^6H_{11/2}$	0.0923	0.0366	0.6410
$^6H_{9/2}$	0	0.0176	0.1985
$^6F_{11/2}$	0.9387	0.8292	0.2048
$^6F_{9/2}$	0	0.5736	0.7213
$^6H_{7/2}$	0	0.0007	0.0392
$^6H_{5/2}$	0	0	0.0026
$^6F_{7/2}$	0	0.1360	0.7046
$^6F_{5/2}$	0	0	0.3452
$^6F_{3/2}$	0	0	0.0610
$^6F_{1/2}$	0	0	0
$^4F_{9/2}$	0	0.0047	0.0295
$^4I_{15/2}$	0.0073	0.0003	0.0654
$^4G_{11/2}$	0.0004	0.0145	0.0003
$^4F_{7/2}$	0	0.0768	0.0263

Reduced matrix elements $\left|\langle S, L, J|U^\lambda|S', L', J'\rangle\right|^2$ for Ho^{3+} (aquo), being $^{2S+1}L_J = {}^5I_8$ the ground state. The wavelength is given in nm [2].

Upper state	$\lambda = 2$	$\lambda = 4$	$\lambda = 6$
$^5I_7$	0.0250	0.1344	1.5216
$^5I_6$	0.0084	0.0386	0.6921
$^5I_5$	0	0.0100	0.0936

(continued)

(continued)

Upper state	$\lambda = 2$	$\lambda = 4$	$\lambda = 6$
$^5I_4$	0	0	0.0077
$^5F_5$	0	0.4250	0.5687
$^5S_2$	0	0	0.2268
$^5F_4$	0	0.2392	0.7071
$^5F_3$	0	0	0.3460
$^3K_8$	0	0	0.1921
$^5G_6$	0.0208	0.0334	0.1578
$^5F_1$	1.5201	0.8410	0.1411
$^5G_5$	0	0	0
$^5G_4$	0	0.5338	0.0002
$^3K_7$	0	0.0315	0.0359
$^3H_5$	0.0058	0.0046	0.0338
$^3H_6$	0	0.0790	0.1610

Reduced matrix elements $\left|\langle S, L, J|U^{\lambda}|S', L', J'\rangle\right|^2$ for Er^{3+} (aquo), being $^{2S+1}L_J$ = $^4I_{15/2}$ the ground state. The wavelength is given in nm [1].

Upper state	$\lambda = 2$	$\lambda = 4$	$\lambda = 6$
$^4I_{13/2}$	0.0195	0.1173	1.4299
$^4I_{11/2}$	0.0291	0.0004	0.3969
$^4I_{9/2}$	0	0.1856	0.0122
$^4F_{9/2}$	0	0.5275	0.4612
$^4S_{3/2}$	0	0	0.2230
$^2H_{11/2}$	0.7326	0.4222	0.0927
$^4F_{7/2}$	0	0.1467	0.6280
$^4F_{5/2}$	0	0	0.2237
$^4F_{3/2}$	0	0	0.1256
$^4H_{9/2}$	0	0.0157	0.2278
$^4G_{11/2}$	0.8970	0.5123	0.1172
$^4G_{9/2}$	0.0229	0.0043	0.0794

Reduced matrix elements $\left|\langle S, L, J|U^{\lambda}|S', L', J'\rangle\right|^2$ for Tm^{3+} (aquo), being $^{2S+1}L_J$ = $^3H_6$ the ground state. The wavelength is given in nm [1].

Upper state	$\lambda = 2$	$\lambda = 4$	$\lambda = 6$
$^3F_4$	0.5470	0.7355	0.2462
$^3H_5$	0.1073	0.2323	0.6381
$^3H_4$	0.2300	0.1032	0.5880

(continued)

(continued)

Upper state	λ = 2	λ = 4	λ = 6
$^3F_3$	0	0.3162	0.8408
$^3F_2$	0	0	0.2609
$^1G_4$	0.0006	0.0355	0.2082
$^1D_2$	0	0.3352	0.0919

References

1. W.T. Carnall, P.R. Fields, B.G. Wybourne, J. Chem. Phys. **42**, 3797 (1965).
2. W.T. Carnall, P.R. Fields, K. Rajnak, J. Chem. Phys. **49**, 4424 (1968).
3. W.T. Carnall, P.R. Fields, K. Rajnak, J. Chem. Phys. **49**, 4450 (1968).
4. W.T. Carnall, P.R. Fields, K. Rajnak, J. Chem. Phys. **49**, 4447 (1968).

Index

G. Lozano C. et al., *A Brief Guide to Optical Spectroscopy of Trivalent Rare Earth Ions in Glasses*, Springer Series in Optical Sciences 247,
https://doi.org/10.1007/978-3-032-18501-3

The manufacturer's authorised representative in the EU is Springer Nature Customer Service Centre GmbH, Europaplatz 3, 69115 Heidelberg, Germany. If you have any concerns regarding our products, please contact ProductSafety@springernature.com

Printed and bound by CPI Group (UK) Ltd, Croydon, CR0 4YY
05/10/2026
02228059-0004